Memoirs of the American Mathematical Society

Number 201

John W. Morgan

A product formula
for surgery obstructions

Published by the
AMERICAN MATHEMATICAL SOCIETY
Providence, Rhode Island

VOLUME 14 · NUMBER 201 (first of 2 numbers) · MAY 1978

Abstract: Given a degree one normal map and a closed manifold one can form their product. The result is another degree one normal map. In this paper we calculate the surgery obstruction of the product in terms of the obstruction of the original normal map and invariants of the closed manifold provided that the latter is simply connected. The result depends on the congruence class modulo 4 of the dimension of the closed manifold. Let us call that dimension d. If d is congruent to 2 or 3 modulo 4 then the resulting product has zero surgery obstruction. If d is congruent to 0 modulo 4 then the obstruction of the product is the original obstruction multiplied by the index of the closed manifold. If d is congruent to 1 modulo 4 then the obstruction of the product depends only on the original obstruction and on the de Rham invariant of the closed manifold. We give an example of the last type where the resulting product obstruction is non zero.

AMS (MOS) subject classification (1970):   57D65.

Key words and phrases: Surgery obstruction, Product formula, de Rham invariant.

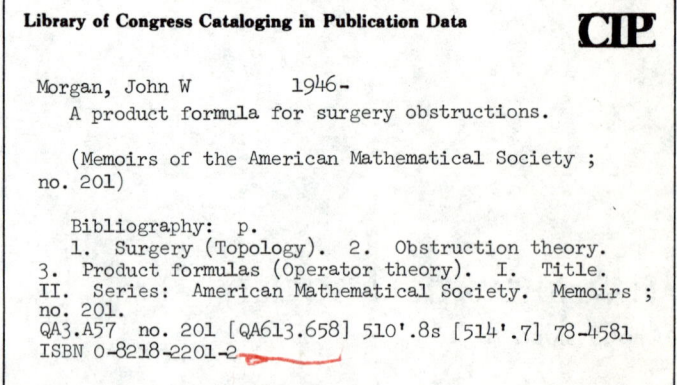

Library of Congress Cataloging in Publication Data   **CIP**

Morgan, John W   1946-
   A product formula for surgery obstructions.

   (Memoirs of the American Mathematical Society ;
no. 201)

   Bibliography: p.
   1. Surgery (Topology). 2. Obstruction theory.
3. Product formulas (Operator theory). I. Title.
II. Series: American Mathematical Society. Memoirs ;
no. 201.
QA3.A57 no. 201 [QA613.658] 510'.8s [514'.7] 78-4581
ISBN 0-8218-2201-2

Table of Contents

The first product formula for degree one normal maps was used by Sullivan in his analysis of the Hauptvermutung for simply connected manifolds. He wanted to treat the Kervaire invariant for normal maps between two dimensional manifolds by high dimensional techniques. To do this he proved, using a geometric argument, that crossing with $\mathbb{C}P^2$ preserves the Kervaire invariant. The analogous statement is obvious for the signature obstruction. This four-fold "geometric periodicity" formed in fact, one of the cornerstones for his geometric analysis of the homotopy theory of the "universal classifying space" in surgery theory, G/PL. Wall generalized this to normal maps between nonsimply connected manifolds. He showed, by a geometric argument, that crossing with $\mathbb{C}P^2$ preserves the nonsimply connected surgery obstruction, [15].

The first general product formula is a consequence of the multiplicativity of the signature for closed, oriented manifolds:

$$I(M \times L) = I(M) \cdot I(L).$$

From this it follows that if $f: M^n \to N^n$ is a degree one normal map with N simply connected, and if L^{ℓ} is a closed, oriented, simply connected manifold, $n + \ell \equiv 0(4)$, then

(∗)   (surgery obstruction of $f \times 1_L: M \times L \to N \times L$) =

(surgery obstruction of f)$\cdot I(L)$

(The surgery obstruction of $f \times 1_L$ is $\frac{1}{8}[I(M \times L) - I(N \times L)]$.)   (∗) is also valid for $n + \ell \equiv 2(4)$ where the surgery obstruction in dimension $4k + 2$ is the Kervaire invariant [11]. There are essentially two types of proof of this formula (as well as all other product formulae): a geometric proof and a homotopy theoretic proof.

The geometric one has its origins in the idea of Rourke and Sullivan [11] that one can give an a-priori, geometric calculation of the Kervaire obstruction for a normal map. A-priori means calculating the obstruction before doing surgery. Geometric means using special cycle representatives for homology classes to calculate the "extra quadratic information" on the middle dimensional homology,which in turn algebraically determines the Kervaire obstruction. In their case, the special cycles are manifolds of one half the dimension of the normal map, which are immersed by using the bundle information of the normal map. The $\mathbb{Z}/2$-quadratic form is the number of double points of the immersion. One then proves the product formula for the Kervaire invariant of $f \times 1_L : M \times L \to N \times L$ by using product cycles in $M \times L$ and examining their double points. Thus the problem is reduced to that of proving a "local product formula" for the product of immersed cycles.

The next child in the lineage of geometric product formulae is the calculation by Sullivan and the author, [9], of the product for normal maps between $\mathbb{Z}/k$-manifolds. The ideas and techniques follow those of the Rourke-Sullivan closely. The crucial case is odd dimensional. The a-priori, geometric information provided in the odd dimensions by the bundle map is a set of embedded manifolds equipped with nowhere zero normal fields. Using these one obtains a "quadratic refinement" of the linking pairing on the torsion subgroups. This refinement determines, in a purely algebraic manner, the surgery obstruction. In a product situation, the problem is again reduced to that of proving a "local product formula" (but this time, for the product of an immersed manifold with an embedded mani-fold with normal field).

In both these situations one makes use of two fortuitous facts. First, all the algebraic information needed is quadratic in nature and therefore subtle only at the prime 2, whereas it is only at the prime 2 (i.e. only after ignoring odd multiples) that one can assume that all cycles are singular submanifolds. Secondly, the structure theorem for finitely generated abelian groups allows one to control the effect on integral homology of surgery.

The main result of this paper is to calculate the surgery obstruction of the product of a non-simply connected normal map and a closed, simply connected manifold. If $f: M^n \to N^n$ is a degree one normal map with surgery obstruction $\sigma(f) \in L_n(\pi_1(N))$, and if L^ℓ is a closed, oriented, simply connected manifold, then $\sigma(f \times 1_L) \in L_{n+\ell}(\pi_1(N))$ is calculated in terms of $\sigma(f)$ and homological invariants of L^ℓ. The idea is to use the bundle information, as in the two previous instances, to give quadratic refinements of the linking and intersection pairings (this time over the fundamental group) and to use these refinements to furnish an a-priori, geometric calculation of the surgery obstructions. The fact mentioned above concerning quadratic forms, the prime 2, and representability of homology classes localized at 2 by manifolds is germane to this discussion, too. However, the results concerning the effects of surgery on the kernel groups have no general analogue for nonsimply connected normal maps. This makes a general a-priori definition of the surgery obstruction difficult. The problem is avoided by assuming that all the kernel modules have the abstract properties of abelian groups. On the basis of this assumption, we give an a-priori, geometric description of the algebraic forms which determine the surgery obstruction.

While this assumption does not hold in general, any product situation can be tailored into a product situation in which it is valid. Once we have the description, we need only prove the "local product formula" which evaluates the forms on product cycles in $M \times L$ in terms of invariants of its factors. Our results are summarized in the following theorem.

Theorem: Let L be a closed, oriented, simply connected manifold, and let $f: M^n \to N^n$ be a degree one normal map with $f/\partial M: \partial M \to \partial N$ a homotopy equivalence. Denote by $I(L^\ell)$ the signature of L if $\ell \equiv 0(4)$, and by $d(L^\ell)$ the de Rham invariant if $\ell \equiv 1(4)$. These are both invariants of the homology of L together with its Poincare duality pairings. Denote by

$\sigma(f)$ the surgery obstruction of f in $L_n(\pi_1(N))$, and by $\sigma(f \times 1_L)$ the surgery obstruction of $f \times 1_L$ in $L_{n+\ell}(\pi_1(N))$. Then

$$\sigma(f \times 1_L) = \begin{cases} 0 & \text{for} \quad \ell \equiv 2,\ 3(4) \\ \sigma(f) \cdot I(L) & \text{for} \quad \ell \equiv 0(4) \\ \varphi(\sigma(f)) \cdot d(L) & \text{for} \quad \ell \equiv 1(4) \end{cases}$$

where $\varphi: L_k(\pi) \to L_{k+1}(\pi)$ is a natural homomorphism for all k and π with $2\varphi(x) = 0$ for all $x \in L_k(\pi)$.

Restated in terms of groups instead of elements, this theorem says that the homomorphism induced by crossing normal maps with manifolds

$$L_n(\pi) \otimes \Omega_\ell \xrightarrow{\omega} L_{n+\ell}(\pi)$$

satisfies

1) $\omega = 0$ if $\ell \equiv 2$ or $3(4)$, and

2) the following diagrams commute

$$
\begin{array}{ccc}
L_n(\pi) \otimes \Omega_{4\ell} & \longrightarrow & L_{n+4\ell}(\pi) \\
\downarrow \text{Id} \otimes \text{signature} & & \downarrow = \\
L_n(\pi) \otimes \mathbb{Z} & \xrightarrow{\ =\ } & L_n(\pi)
\end{array}
$$

$$
\begin{array}{ccc}
L_n(\pi) \otimes \Omega_{4\ell+1} & \longrightarrow & L_{n+4\ell+1}(\pi) \\
\downarrow \text{Id} \otimes \text{de Rham inv.} & & \downarrow = \\
L_n(\pi) \otimes \mathbb{Z}/2 & \xrightarrow{\ \varphi\ } & L_{n+1}(\pi)
\end{array} \quad .
$$

This theorem gives a complete formula for the effect on surgery obstructions of crossing with a simply connected manifold except for the fact that φ is not known in general. It does show, however, that the only possible invariants of the simply connected manifold which can come into play are the signature and the de Rham invariant. If these two invariants vanish, then the surgery obstruction of the product is zero.

φ is known to be zero (i.e. the de Rham invariant has no effect) in many cases. But for $\pi_1(N) = \mathbb{Z}$, with the generator an orientation reversing loop (this situation is denoted $(\mathbb{Z}, -)$), φ induces an isomorphism:

$$\varphi: L_3(\mathbb{Z}, -) \xrightarrow{\cong} L_4(\mathbb{Z}, -).^{[1]}$$

This gives an example in which crossing with an odd dimension manifold does not annihilate the surgery obstruction. The theorem and the calculation of φ are also valid if we begin with a normal map which is a simple homotopy equivalence on the boundary and compute the surgery obstructions in $L_*^s(\pi)$.

The first general product formula of this type was announced by Williamson in [16] for the case of an odd dimensional normal map crossed with an even dimensional manifold. The result is the same as the one we obtain in the case (i.e. multiplication by the signature). It seems from the sketch of the proof given there that Williamson had in mind an argument similar to ours. Using the idea of crossing with S^1, Shaneson, [12], extended Williamson's argument to the case of crossing any normal map with an even dimensional manifold. This gives results only about surgery obstructions in $L_*(\pi)$ not in $L_*^s(\pi)$, however.

This paper can be outlined as follows. Chapters I and II show that the usual analysis of simply connected surgery--using intersection and linking pairings both to study the effect of low dimensional surgery and to find the obstruction to doing middle dimensional surgery--remains valid in the nonsimply connected case, if the kernel modules have the abstract properties of abelian groups. The remainder of the paper demonstrates that in the product situation the kernel modules have these properties. It also evaluates the pairings which determine the surgery obstruction.

---

[1] This calculation is a reinterpretation of the connection between the Kervaire invariant and the signature described in [9], see also [8].

More explicitly, section I.1 sets up the notation (we follow Wall's notation, [15]), and recapitulates some of the fundamental theory of [15] which we will use.

Section I.2 introduces the assumption that all the kernel modules have the abstract properties of abelian groups. Normal maps with this type of kernel modules are called nice, normal maps. We prove that the short exact sequence of the usual universal coefficient theorem is valid in the context of such kernel modules. From that we prove that there are non-singular linking and intersection pairings on the kernel modules, as in the simply connected case, which capture all of Poincare duality. These pairings form the basis for the analysis of the effect of doing surgery. All is the complete analogue of the simply connected case.

In section I.3, we calculate the effect of low dimensional surgery on a nice normal map. Here we generalize the theorem in the simply connected case that one can do the surgery until the only non zero kernel module is in the "middle dimension" <u>while keeping track of the kernel</u> <u>modules, their linking and intersection pairings, and Λ-bases.</u> In the odd dimensional case, we are able to replace a nice, normal map $f: M^{2n-1} \to N^{2n-1}$, up to normal bordism, by $f': M' \to N$ such that

$$K_i(f') = \begin{cases} 0 & i \neq n \\ \mathbf{Tor}\, K_n(f) & i = n \end{cases}$$

and such that the self-linking on $\mathbf{Tor}\, K_n(f)$ is unchanged. In the even dimensional case, we replace $f: M^{2n} \to N^{2n}$ by $f': M' \to N$ such that

$$K_i(f') = \begin{cases} 0 & i \neq n \\ K_n(f)/\mathrm{Tor} & i = n \end{cases}$$

The intersection pairing on $K_n(f)/\mathrm{Tor}$ is the original one.

In section II.1 we use the analysis in chapter I and the Rourke-

Sullivan immersed cycle idea to give a geometric, a-priori definition of the surgery obstruction in $L_{2n}(\pi)$ (or $L_{2n}^s(\pi)$) for an even dimensional, nice, normal map f: $M^{2n} \to N^{2n}$. The idea is to define the self-intersection form, μ_f, on $K_n(f)$ thus producing a triple $(K_n(f)/\text{Tor}, \lambda, \mu_f)$ <u>before</u> doing the low dimensional surgery. The element this triple determines in $L_{2n}(\pi)$ (or $L_{2n}^s(\pi)$) is the Wall surgery obstruction for f. To prove this, we use the results of section I.3 to replace f by an f' with $\sigma(f) = \sigma(f')$ and $K_n(f') = K_n(f)/\text{Tor}$ without disturbing the λ or μ-form. The argument is completed by showing that μ_f on $K_n(f)/\text{Tor}$ agrees with Wall's μ-form for $K_n(f')$.

In section II.2 we give the a-priori definition of the surgery obstruction in the odd dimensions. Here our formalism differs from Wall's. Associated to a nice normal map, f: $M^{2n-1} \to N^{2n-1}$, we find a triple $\{\text{Tor } K_{n-1}(f), \ell, q_f\}$; ℓ is the non singular linking pairing, ℓ: Tor $K_{n-1}(f) \times$ Tor $K_{n-1}(f) \to \mathbb{Q}/\mathbb{Z} \otimes \Lambda$, and q_f is an a-priori, geometrically defined quadratic refinement of ℓ. q_f is algebraically determined on the odd torsion by ℓ. On the two-torsion, we use cycles which are embedded submanifolds with nowhere zero normal fields coming from the bundle data. The normal field allows us to push the submanifolds off themselves to gain the "extra factor of 2" required to define q_f. We then sketch a proof that this triple algebraically determines the Wall surgery obstruction--though we do not make use of this in the sequel.

In section II.3 we discuss when the triples defined in II.1 and II.2 determine the zero surgery obstruction in $L_n(\pi)$ (or $L_n^s(\pi)$). For the even dimensional case (G, λ, μ) determines 0 in $L_{2n}(\pi)$ if and only if there is a "subkernel" $K \subset G$. (That is a submodule K of G with $\lambda/K \times K = 0$, $\mu/K = 0$, and $\text{Ad}(\lambda)K \to (G/K)*$ an isomorphism.) This is just Wall's condition. For the odd dimensional case, (Tor K_{n-1}, ℓ, q_f) determines 0 in $L_{2n-1}(\pi)$, if and only if there is a resolution

$$0 \longrightarrow A_{n-1} \longrightarrow F_{n-1} \longrightarrow \text{Tor } K_{n-1}(f) \longrightarrow 0 \ ,$$

with A_{n-1} and F_{n-1} free Λ-modules, and a pairing

$$I: F_{n-1} \times F_{n-1} \longrightarrow \mathbb{Q} \otimes \Lambda$$

which "lifts" ℓ and q_f in an appropriate sense, so that

$$\text{Ad}(I): A_{n-1} \longrightarrow F_{n-1}{}^*$$

is an isomorphism.

In chapter III, we change from non-simply connected surgery theory to oriented, closed, simply connected manifolds. We study algebraic invariants of their Poincare duality intersection and linking pairings. Two invariants interest us: the signature, in the case of symmetric intersection pairings (4k-manifolds), and the de Rham invariant, in the case of skew-symmetric linking pairings (4k + 1-manifolds).

In chapter IV we give the proof of our product formula. In section IV.1 we show that if $f: M^{2n} \to N^{2n}$ is a degree one normal map, then

$$\sigma(f \times 1_L \ell) = \begin{cases} 0 & \text{for} \quad \ell \equiv 2,3\,(4) \\ \sigma(f) \cdot I(L) & \text{for} \quad \ell \equiv 0\,(4) \\ 0 & \text{for} \quad \ell \equiv 1\,(4) \quad \text{if} \quad d(L) = 0. \end{cases}$$

First we establish a "local product formula" which evaluates the pairings and their quadratic refinements in a product situation. Then, the results of chapters I and II are applied to this formula in order to prove the above.

We can assume f is n-connected and that $K_n(f)$ is a free Λ-module with a non-singular intersection pairing, λ_f, and a μ-form, μ_f. The kernel modules for $f \times 1_L \ell$ are $K_n(f) \otimes H_i(L^\ell)$. The intersection and linking pairings are the tensor product of those on $H_*(L)$ with λ_f. If $\ell \equiv 0\,(2)$, then

$$\mu_{f \times 1_L}(x \otimes y) = \mu_f(x) \cdot (y \cdot y).$$

If $\ell \equiv 3(4)$, or if $\ell \equiv 1(4)$ and $\ell(y,y) = 0$, then

$$q_{f \times 1_L}(x \otimes y) = \mu_f(x) \otimes \ell(y,y).$$

This leads to the following two results. If $\ell \equiv 0(2)$ and $I(L^\ell) = 0$, then let $K \subset H_{\ell/2}(L)/\text{Tor}$ be a subkernel. We can use $K_n(f) \otimes K \hookrightarrow K_n(f) \otimes H_{\ell/2}(L)/\text{Tor} = K_{n+\ell/2}(f \times 1_L)/\text{Tor}$ as a subkernel to show $\sigma(f \times 1_L) = 0$. If $\ell \equiv 1(2)$ and $d(L) = 0$, then there is a product resolution for $\text{Tor } K_{n+(\ell-1)/2}(f \times 1_L)$ $(\cong K_n(f) \otimes \text{Tor } H_{(\ell-1)/2}(L^\ell))$. This product resolution admits a pairing into $\mathbb{Q} \otimes \Lambda$ which "lifts" the linking pairing and its quadratic refinement, and is non-singular. This proves $\sigma(f \times 1_L) = 0$. There is a slight twist here in that the pairing is not always the tensor product of λ_f with a pairing on a resolution for $\text{Tor } H_{(\ell-1)/2}(L)$ in case $\ell \equiv 1(4)$. This is a consequence of the fact that $\ell(y,y)$ is not identically zero in this case. When $d(L) = 0$ though, $\ell(y,y)$ is zero for enough elements y (for at least $\frac{1}{2}$ of some generating set) to allow a proof that the pairing is non-singular. By a simple additivity argument, we go from the above two results to the statement of the main theorem in the case of an even dimensional normal map crossed with any closed, orineted, simply connected manifold.

The case of an odd dimensional normal crossed with any closed, oriented, simply connected manifold is dealt with in sections IV.2 and IV.3. When considering an odd dimensional normal map, we first do surgery until $f: M^{2n-1} \to N^{2n-1}$ is $(n-1)$ connected. Then, we cut out a regular neighborhood, U, of a union of spheres $\{S_i^{n-1}\}$ which generate $K_{n-1}(f)$. We can assume that U is homotopy equivalent to a wedge of spheres and that $f|U: U \to D^{2n-1} \hookrightarrow N$. Let M_0 be $\overline{M - U}$, and let N_0 be $N - D$. All the kernel modules for $f: M_0 \to N_0$ are zero except for

$$0 \longrightarrow K_n(M_0, \partial M_0) \longrightarrow K_{n-1}(\partial M_0) \longrightarrow K_{n-1}(M_0) \longrightarrow 0,$$

which are all free based Λ-modules. We let L^ℓ be a simply connected manifold whose signature or de Rham invariant is 0, and we form the product
normal map of pairs $M_0 \times L \to N_0 \times L$. This map is not a homotopy equivalence on the boundary. In terms of either a resolution for
Tor $H_{(\ell-1)/2}(L^\ell)$ with its non-singular pairing, or a subkernel in $H_{\ell/2}(L^\ell)$,
we find a <u>canonical</u> normal bordism, W, of the normal map on the boundary
to a simple homotopy equivalence. We study the question of doing surgery
on $M_0 \times L \cup W \to N_0 \times L$.

Now we have a normal map which is a homotopy equivalence on the
boundary. A study of the kernel modules of this normal map reveals that
they are products of $K_n(M_0, \partial)$ or $K_{n-1}(M_0)$ with groups associated to the
homology of L. The problem is once again reduced to a "local product
formula". This time, however, things are complicated somewhat by the
fact that W is not a product. We show, in the end, that it is always
possible to do surgery on this map to make it a homotopy equivalence of
pairs. This leaves the "other side",

$$W \cup U \times L \longrightarrow D^{2n-1} \times L.$$

It is a normal map into a simply connected manifold and is a homotopy equivalence on the boundary. If $(2n - 1 + \ell)$ is odd, then we can automatically
do surgery on this normal map relative to its boundary. If $(2n - 1 + \ell)$ is
even, then the only obstruction is the signature or Kervaire invariant.
We identify this obstruction with the signature or Kervarie invariant of
$M \times L \to N \times L$. The latter vanishes by the product formula for these
invariants.

In chapter V we consider the example of normal maps of manifolds with
fundamental group $\mathbb{Z}$ and generator orientation reversing. We identify
$L_3(\mathbb{Z}, -)$ with $\mathbb{Z}/2$ by the Kervaire invariant along a codimension one sub-

manifold dual to the generator of π_1. We identify $L_4(\mathbb{Z}, -)$ with $\mathbb{Z}/2$ as follows. Take a codimension one submanifold dual to the generator of π_1 and make the normal map a homotopy equivalence there. Then cut the map open along this manifold to obtain a normal map between oriented manifolds which is a homotopy equivalence on the boundary. Take the signature obstruction of this and reduce modulo 2. With these descriptions the connection between the Kervaire invariant and signature of [9] is easily translated to: $\varphi: L_3(\mathbb{Z}, -) \to L_4(\mathbb{Z}, -)$ is an isomorphism.

The author wishes to thank L'institut des Hautes Etudes Scientifiques for their hospitality during the preparation of this paper, especially Mme Martin and Mme Cabanes for their expert assistance and typing, and also Ms. Kate March who prepared the final manuscript for publication.

A PRODUCT FORMULA FOR SURGERY OBSTRUCTIONS

CHAPTER I: Preliminaries

<u>Section I.1 - Notation</u>. Throughout this paper we use the notation and
results of [15]. The purpose of this section is to set up this notation
and outline some of the more important results which we rely upon. A
degree one, normal map or, normal map for short, is a diagram

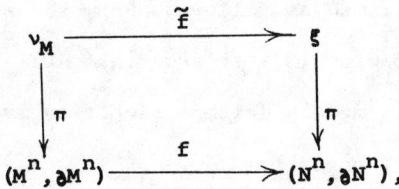

and is denoted by $(f,\tilde{f})$: $(M,\partial M) \to (N,\partial N)$ or by f: M → N. Here M and N
are manifolds; f is a degree one map of pairs; and $\tilde{f}$ is a bundle map
covering f. This forces ξ to be fiber homotopy equivalent to ν_N. A
normal bordism between two normal maps is just a normal map into N × I
which on the two ends is the two given normal maps.

There are algebraically defined groups, $L_n^s(\pi)$ and $L_n(\pi)$, depending
only on π and the residue of n modulo 4. If f: $M^n \to N^n$ is a simple
homotopy equivalence (or homotopy equivalence) on the boundary, then
there is the surgery obstruction for f, $\sigma(f) \in L_n^s(\pi_1(N))$ (or
$\sigma(f) \in L_n(\pi_1(N))$). This element is the only obstruction to replacing f
by a map f' which is normally bordant to f relative ∂M and which is
a simple homotopy equivalence of pairs (a homotopy equivalence of pairs),
for $n \geq 5$. Furthermore, all elements in $L_n^s(\pi)$ (or $L_n(\pi)$) are surgery

Received by the editor May 23, 1977 and, in revised form, Nov. 21, 1977.
The author was partially supported by the Sloan Foundation and NSF grant
#MCS76 08230.

obstructions of a normal map. Thus, to study $L_n^s(\pi)$ or $L_n(\pi)$ it suffices to study normal maps in dimensions greater than or equal to 5, and conversely. Throughout this paper we assume that all our normal maps are of dimension at least 5. The process of building the normal bordisms from f to a simple homotopy equivalence, or, in fact, any more highly connected normal map is surgery.

We denote the integral group ring $\mathbb{Z}[\pi]$ by Λ. We equip π with an orientation homomorphism w: $\pi \to \{\pm 1\}$, and Λ with a canonical anti-involution, $\alpha \to \bar{\alpha}$, which sends $\Sigma\ n_i g_i \to \Sigma\ w(g_i) \cdot n_i\ g_i^{-1}$. For n = 2k, $L_n(\pi)$ and $L_n^s(\pi)$ are defined as follows. Form the semi-group under orthogonal direct sum of triples (G,λ,μ) such that

1) G is a free Λ-module (with a simple equivalence class of bases in the case of $L_{2k}^s(\pi)$);

2) λ: $G \times G \to \Lambda$ is

a) Λ-linear in the second variable,

b) $\lambda(x,y) = (-1)^k\ \overline{\lambda(y,x)}$, and

c) $ad(\lambda)$: $G \to Hom_\Lambda(G,\Lambda)$ is an isomorphism (simple isomorphism in the case of $L_{2k}^s(\pi)$);

3) μ: $G \to Q_k = \Lambda/\{\nu - (-1)^k\ \bar{\nu}\}$ satisfies

a) $\mu(x) + (-1)^k\ \overline{\mu(x)} = \lambda(x,x)$ in Λ,

b) $\mu(x + y) - \mu(x) - \mu(y) = [\lambda(x,y)]$ in Q_k, and

c) $\mu(xa) = \bar{a}\ \mu(x)a$ for $a \in \Lambda$.

$L_{2k}(\pi)$ or $L_{2k}^s(\pi)$ is the associated Grothendieck group modulo one relation: hyperbolic forms are set equal to zero. A triple (G,λ,μ) is hyperbolic if there is an isomorphism from it to a direct sum of copies of

$$(\Lambda \oplus \Lambda \text{ with basis } \{x,y\},\ \lambda = \begin{pmatrix} 0 & 1 \\ (-1)^k & 0 \end{pmatrix},\ \mu(x) = \mu(y) = 0).$$

In the case of $L_{2k}^s(\pi)$ this isomorphism must be a simple isomorphism of based Λ-modules. Wall proves that a triple (G,λ,μ) determines 0 in $L_{2k}(\pi)$ if and only if it has a subkernel. A subkernel is a free submodule $K \subset G$ on which λ and μ vanish identically and such that $\text{ad}(\lambda)\colon K \to \text{Hom}(G/K,\Lambda)$ is an isomorphism. In the case of $L_{2k}^{s}(\pi)$, K must be equipped with a basis so that $\text{ad}(\lambda)$ is a simple isomorphism of based Λ-modules. Given such a subkernel, K, (G,λ,μ) is actually isomorphic to a hyperbolic form.

A normal map $f\colon M^{2k} \to N^{2k}$ determines an element in $L_{2k}(\pi_1(N))$ if $f|\partial M$ is a homotopy equivalence, and in $L_{2k}^{s}(\pi_1(N))$ if $f|\partial M$ is a simple homotopy equivalence.

If f is a normal map, then $f_{*}\colon H_i(M;\Lambda) \to H_i(N;\Lambda)$ is always onto. We denote the kernel of f_{*} on H_i by $K_i(f)$. If $f|\partial M$ is a homotopy equivalence, then one can do surgery to make $K_i(f) = 0$ for $i \neq k$ (i.e. $f_{*}\colon H_i(M;\Lambda) \to H_i(N;\Lambda)$ an isomorphism for $i \neq k$). Furthermore, one can make $K_k(f)$ a free Λ-module. $\lambda\colon K_k(f) \times K_k(f) \to \Lambda$ is the usual intersection pairing (over the fundamental group) of middle dimensional cycles. To define $\lambda(x,y)$ one takes based, oriented, simply connected cycles, $X^k \to M$ and $Y^k \to M$, representing x and y. By a slight shift of Y^k we can insure that X^k and Y^k intersect only in isolated points where two k-simplices meet transversally. Associated to such a point p is $\epsilon_p g_p{}'$ with $\epsilon_p = \pm 1$ and $g_p \in \pi_1$. The ϵ_p measures the usual compatibility of orientations when the local orientation at the base point is moved out along a path in Y^k to p, whereas g_p is the class of any loop beginning at the base point traveling in X to p and returning in Y to the base point. $\lambda(x,y) = \sum\limits_{\substack{\text{points of} \\ \text{intersection, } p}} \epsilon_p g_p$. To define μ, the self-intersection form, we use the bundle map $\tilde{f}$ covering f to immerse a sphere representing any $x \in K_k(f)$, $S^k \looparrowright M^{2k}$. Then $\mu(x)$ is the self intersection of this sphere. It is defined by making the immersion have only transversal double points, associating to each double point $[\epsilon_p g_p] \in Q_k$, and adding

over the double points. The group element g_p is defined by starting at
the base point, moving out along the S^k to the double point, switching
sheets, and coming back to the base point along S^k. The ϵ_p measures the
sign of the two oriented, intersecting sheets when the local orientation
at the base point is pushed out to the double point along the second half
of the path.

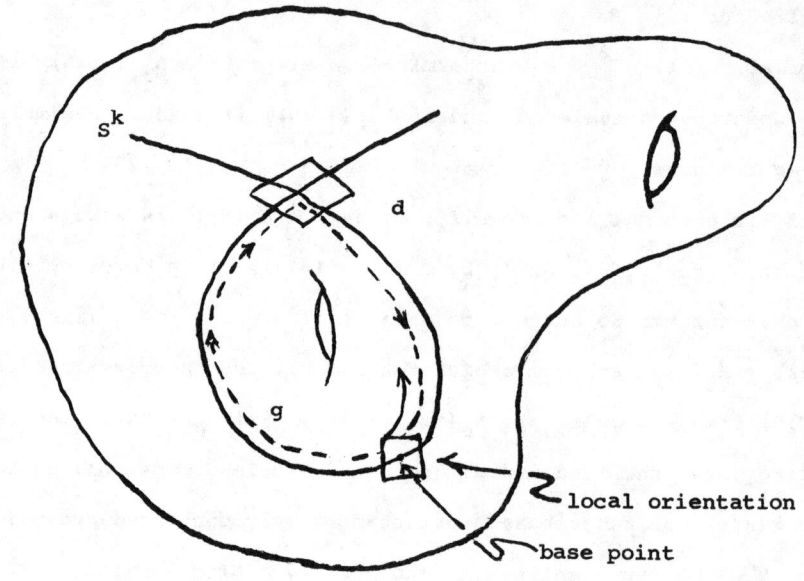

If $f|\partial M$ is a simple homotopy equivalence, then the chain complex
for f, $C_*(f)$, has a geometric basis. Poincaré duality is a simple isomorphism in this basis. If $K_i(f) = 0$ for $i \neq k$, and $K_k(f)$ is free, then
$C_*(f)$ induces a simple equivalence class of basis on $K_k(f)$, (the one with
the property that choosing cycles representatives for the basis defines
a simple chain homotopy equivalence $K_k(f) \to C_*(f)$). With this basis on
$K_k(f)$, the adjoint of λ is a simple homotopy equivalence
$K_k(f) \to \mathrm{Hom}_\Lambda(K_k(f),\Lambda) = K_k(f)^*$. The class in $L_{2k}(\pi)$ of the triple
$(K_k(f),\lambda,\mu)$ is the surgery obstruction of f, if $f|\partial M$ is a homotopy
equivalence. Equipping $K_k(f)$ with its basis and taking the triple in
$L_{2k}^s(\pi)$ gives the surgery obstruction, if $f|\partial M$ is a simple homotopy

equivalence.

For $n = 2k + 1$, $L_n(\pi)$ and $L_n^s(\pi)$ are defined by considering subkernels in the standard hyperbolic form. Let H_t denote the hyperbolic form:

1) $\oplus \Lambda$ with basis $\{e_1, \ldots, e_t, f_1, \ldots, f_t\}$;

2) $\lambda(e_i, e_j) = 0 = \lambda(f_i, f_j)$, $\lambda(e_i, f_j) = \delta_{ij}$, and $\lambda(f_i, e_j) = (-1)^k \delta_{ij}$,

3) $\mu(e_i) = 0 = \mu(f_i)$.

Then, an element in $L_n(\pi)$ is determined by a subkernel $K \subset H_t$, and an element in $L_n^s(\pi)$ is determined by a based subkernel in H_t. The algebraic relations which tell when a subkernel determines the zero element in $L_n(\pi)$ or $L_n^s(\pi)$ are complicated and do not concern us here.

A normal map $f: M^{2k-1} \to N^{2k+1}$ determines an element in $L_{2k+1}(\pi)$ or $L_{2k+1}^s(\pi)$ as follows. We can do surgery until $K_i(f) = 0$ for $i < k$. Let $\{s_1^k, \ldots, s_t^k\}$ be disjointly embedded k-spheres representing a generating set for $K_k(f)$. We can assume that f maps tubular neighborhoods of these spheres to a fixed disk in N. Cut out the union of the interiors of these tubes (denoted U) from M, as well as the disk from N. Denote by $M_0^{2k+1} \xrightarrow{f|} N_0^{2k+1}$ the normal map restricted to the complements. The bundle map $\tilde{f}$ provides trivializations of the normal tubes around the s_i^k's. Thus, it gives a homeomorphism $\partial U \cong S^k \times S^k$, and hence, a natural geometric Λ-basis for $K_k(f|\partial U) = K_k(f|\partial M_0)$. This basis gives an isomorphism of $(K_k(f|\partial M_0), \lambda, \mu)$ with the hyperbolic triple, H_t. The image $0 \to K_{k+1}(f|M_0, f|\partial M_0) \xrightarrow{\partial} K_k(f|\partial M_0)$ is a subkernel if $f|\partial M$ is a homotopy equivalence, and is a based subkernel if $f|\partial M$ is a simple homotopy equivalence. The basis again comes from the geometric basis for $C_*(f|M_0)$. This is the subkernel, or based subkernel, whose class in $L_{2k+1}(\pi)$ or $L_{2k+1}^s(\pi)$ is the surgery obstruction $\sigma(f)$.

Section I.2 - Nice, normal maps. The main restriction on the homology

and cohomology groups of a closed, oriented n-manifold M^n is that they

must satisfy Poincare duality:

$$H^i(M;\mathbb{Z}) \xrightarrow[\cong]{\cap[M]} H_{n-i}(M;\mathbb{Z}) .$$

This duality isomorphism can be re-interpreted via the universal coeffi-

cient theorem completely in terms of the homology groups. It says:

1)   there are non-singular "intersection pairings"

$$(H_i(M;\mathbb{Z})/\text{Torsion}) \otimes (H_{n-i}(M;\mathbb{Z})/\text{Torsion}) \longrightarrow \mathbb{Z}, \quad \text{and}$$

2)   there are non-singular "linking pairings"

$$\text{Torsion } H_i(M;\mathbb{Z}) \otimes \text{Torsion } H_{n-i-1}(M;\mathbb{Z}) \longrightarrow \mathbb{Q}/\mathbb{Z} .$$

(Non-singular means, in each case, that the associated adjoints to these

pairings are isomorphisms.)

If $f: M^n \to N^n$ is a degree 1 map between oriented manifolds with

$f|\partial M^n: \partial M^n \to \partial N^n$ a homotopy equivalence, then $f_*: H_i(M;\mathbb{Z}) \to H_i(N;\mathbb{Z})$

is onto and we denote by $K_i(f)$ the kernel of this homomorphism. Likewise

$f^*: H^i(N;\mathbb{Z}) \to H^i(M;\mathbb{Z})$ is injective, and we denote its cokernel by $K^i(f)$.

Poincare duality holds also for these groups, and again it can be re-

interpreted to say that the groups $K_i(f)$ admit non-singular linking and

intersection pairings. In fact, these pairings play a central role in

the analysis of simply-connected surgery, see [6].

In this section, we will develop an analogue of this for the kernel

groups of a degree one normal map between non-simply connected manifolds

with coefficients in the group ring of the fundamental group, $\Lambda = \mathbb{Z}[\pi_1]$.

The first problem is that of the structure of the underlying kernel

modules, $K_i(f;\Lambda) = \text{Ker } f_*: H_i(M;\Lambda) \to H_i(N;\Lambda)$. These can have complicated

structure as Λ-modules. We solve this problem by assuming all kernel modules in question have all the abstract, algebraic properties of finitely generated abelian groups. In particular we assume that $K_i(f;\Lambda)$, denoted $K_i(f)$, is isomorphic to $F \underset{i}{\oplus} \Lambda/m_i\Lambda$ with $m_i \in \mathbb{Z}^+$, and F a finitely generated free Λ-module. The torsion subgroup, Tor $K_i(f)$, is $\underset{i}{\oplus} \Lambda/m_i\Lambda$. The quotient $K_i(f;\Lambda)/\text{Tor } K_i(f)$ is a free Λ-module. The main result of this section is that for degree one normal maps with such kernel modules there are non-singular intersection and linking pairings generalizing (1) and (2) above. This is proved by establishing the universal coefficient theorem short exact sequence relating $K_i(f)$ and $K^i(f)$. In calculating obstructions to performing surgery to produce a simple homotopy equivalence, it is necessary to work with Λ-based chain complexes. At the end of this section we discuss chain complex models for the chains representing the kernel models.

Definition I.2.1: K is <u>a nice</u> Λ-<u>module</u> if and only if it is isomorphic to a finite direct sum of copies of Λ and copies of $\Lambda/n_i\Lambda$ for $n_i \in \mathbb{Z}^+$.

One type of nice Λ-module is $F \underset{\Lambda}{\underset{\mathbb{Z}}{\otimes}} A$, for F a free finitely generated Λ-module and A a finitely generated abelian group. The torsion subgroup of a nice Λ-module K, denoted Tor K, is the subgroup of $x \in K$ such that $n \cdot x = 0$ for some $n \in \mathbb{Z} - \{0\}$. Tor K is a nice Λ-module, and $K/\text{Tor } K$ is a finitely generated free Λ-module.

Definition I.2.2: A normal map $f: M^n \to N^n$ is <u>a nice normal map</u> if

1) $f_*: \pi_1(M) \to \pi_1(N)$ is an isomorphism,

2) $f|\partial M: \partial M \to \partial N$ is a homotopy equivalence, and

3) $K_i(f)$ is a nice Λ-module for all i.

If $f: M^n \to N^n$ is a normal map satisfying property 1) above, then there are intersection pairings

$$K_i(f) \times K_{n-i}(f) \longrightarrow \Lambda.$$

These are defined by taking based, oriented simply connected cycles X^i and Y^{n-i} in M^n representing $x \in K_i(f)$ and $y \in K_{n-i}(f)$. We make these intersect transversally and count the intersection points in $\mathbb{Z}[\pi]$ exactly as described in section 1. If X^i is the boundary of C^{i+1}, then C^{i+1}, $C^{i+1} \cdot Y^{n-i}$ is a based 1-chain in M^n whose boundary is, on the one hand $X^i \cdot Y^{n-i}$, and on the other 0. This proves that the pairing is well defined, (compare [15], page 45). It satisfies $y \cdot x = (-1)^{i \cdot (n-i)} \overline{(x \cdot y)}$. It is $\mathbb{Z}$-bilinear, Λ-linear in the second variable, and Λ-anti-linear in the first variable. Hence, it induces a Λ-module homomorphism

$$\mathrm{Ad}(\cdot) \colon K_i(f) \longrightarrow \mathrm{Hom}_\Lambda(K_{n-i}(f), \Lambda) = K_{n-i}(f)^*.$$

If $K_i(f)$ and $K_{n-i}(f)$ are nice Λ-modules, then $x \cdot y = 0$ if either $x \in \mathrm{Tor}\, K_i(f)$ or $y \in \mathrm{Tor}\, K_{n-i}(f)$, since Λ has no integral torsion. In this case we consider the intersection pairing as a map

$$K_i(f)/\mathrm{Tor} \times K_{n-i}(f)/\mathrm{Tor} \longrightarrow \Lambda,$$

and its adjoint as a Λ-homomorphism

$$\mathrm{Ad}(\cdot) \colon K_i(f)/\mathrm{Tor} \longrightarrow \mathrm{Hom}_\Lambda(K_{n-i}(f)/\mathrm{Tor}, \Lambda).$$

When the modules $K_i(f)$ are nice, there are linking pairings on the torsion subgroups:

$$\ell \colon \mathrm{Tor}\, K_i(f) \times \mathrm{Tor}\, K_{n-i-1}(f) \longrightarrow \mathbb{Q}/\mathbb{Z} \otimes \Lambda$$

which, as we shall see, are defined similarly. By a resolution of a nice, torsion Λ-module, T, we mean a short exact sequence of free, based Λ-modules:

$$0 \longrightarrow A \overset{\varphi}{\longrightarrow} F \overset{\beta}{\longrightarrow} T \longrightarrow 0,$$

with bases $\{x_1, \ldots, x_k\}$ for A and $\{y_1, \ldots, y_k\}$ for F and such that

$\varphi(x_i) = n_i y_i$ for $n_i \in \mathbb{Z}^+$. Such a resolution is equivalent to giving an isomorphism

$$T \cong \bigoplus_i \Lambda/n_i \Lambda .$$

Given a resolution

$$0 \longrightarrow A_i \xrightarrow{\varphi_i} F_i \xrightarrow{\rho_i} \text{Tor } K_i(f) \longrightarrow 0$$

a <u>chain realization</u> of it is a collection of based cycles $\{\tilde{Y}_j^i\}$ in $\tilde{M}$, the universal cover of M, representing the image under ρ_i of the basis for F_i and based chains, $\{\tilde{C}_j^{i+1}\}$ in $\tilde{M}$ with $\partial \tilde{C}_j = n_j Y_j$. We identify F_i and A_i with the free Λ-modules on the $\{\tilde{Y}_j\}$ and $\{\tilde{C}_j\}$, respectively.

Given resolutions

$$0 \longrightarrow A_i \longrightarrow F_i \longrightarrow \text{Tor } K_i(f) \longrightarrow 0$$

and

$$0 \longrightarrow A_{n-i-1} \longrightarrow F_{n-i-1} \longrightarrow \text{Tor } K_{n-i-1}(f) \longrightarrow 0$$

such that the cycles in question are disjoint in M (which is generically the case), then we can calculate intersections exactly as before to give homomorphisms $I_1 \colon F_i \times A_{n-i-1} \to \Lambda$ and $I_2 \colon A_i \times F_{n-i-1} \to \Lambda$. We define $I \colon F_i \times F_{n-i-1} \to \mathbb{Q} \otimes \Lambda$ by

$$I(x,y) = \frac{1}{m} I_1(x, \varphi_{n-i-1}^{-1}(my)) .$$

Using the intersection of the $(i+1)$-chains and $(n-i)$-chains, we see that

(*) $$I(x,y) = (-1)^i \frac{1}{p} \cdot (I_2(\varphi_i^{-1}(px), y)) .$$

Thus, $I \colon F_i \times F_{n-i-1} \to \mathbb{Q} \otimes \Lambda$ induces a pairing

$$\ell \colon \text{Tor } K_i(f) \times \text{Tor } K_{n-i-1} \longrightarrow \mathbb{Q}/\mathbb{Z} \otimes \Lambda .$$

This map is Λ-linear in the second variable and Λ anti-linear in the first (as are I, I_1, and I_2).

Lemma I.2.2:   a)   The pairing ℓ: Tor $K_i(f) \times$ Tor $K_{n-i-1}(f) \rightarrow \mathbb{Q}/\mathbb{Z} \otimes \Lambda$ is

independent of the resolutions and chain realizations.

b)   Given resolutions and chain realizations for Tor $K_i(f)$ and

Tor $K_{n-i-1}(f)$, $i \neq n-i-1$, it is possible by moving the cycles

of dimension $(n-i-1)$ to realize any pairing

$$I: F_i \times F_{n-i-1} \longrightarrow \mathbb{Q} \otimes \Lambda$$

for which the following hold:

1) I is Λ-linear in the second variable and Λ-anti-linear in

the first,

2) $I|A_i \times F_{n-i-1}$ and $I|F_i \times A_{n-i-1}$ take values in $\mathbb{Z} \otimes \Lambda$, and

3) I induces ℓ.

The proof is a standard exercise in the theory of chains, and is

left to the reader.

Corollary: We have a well-defined linking pairing

$$\ell: \text{Tor } K_i(f) \times \text{Tor } K_{n-i-1}(f) \longrightarrow \mathbb{Q}/\mathbb{Z} \otimes \Lambda,$$

and its adjoint which is a homomorphism

$$\text{ad}(\ell): \text{Tor } K_i(f) \longrightarrow \text{Hom}_\Lambda(\text{Tor } K_{n-i-1}(f), \mathbb{Q}/\mathbb{Z} \otimes \Lambda).$$

We will show that if $f: M^n \rightarrow N^n$ is a nice normal map, then both the

intersection and linking pairings are nonsingular, i.e. their adjoints

are isomorphisms. First, we prove a universal coefficient theorem for

Λ-chain complexes whose homology is nice. This is the analogue of the

usual universal coefficient theorem for chain complexes over the integers.

Adding Poincare duality to this result gives a proof of the non-singular-

tiy of the intersection and linking pairings.

Lemma I.2.3: If $\{C_*, \partial\}$ is a Λ-chain complex with each C_i a free Λ-module,

$C_i = 0$ for $i < 0$, and $H_i(C_*)$ a nice Λ-module for all i, then the modules of cycles $Z_i \subset C_i$ and boundaries $B_{i+1} \subset C_i$ are stably free.

Proof: The proof goes by induction using the short exact sequences

$$0 \longrightarrow Z_i \longrightarrow C_i \longrightarrow B_i \longrightarrow 0 \ ,$$

and

$$0 \longrightarrow B_{i+1} \longrightarrow Z_i \longrightarrow H_i \longrightarrow 0.$$

The only point worth mentioning is that if Z_i is stably free and $H_i = \oplus \Lambda/n_i\Lambda$, $n_i \in \mathbb{Z}$, then we construct

$$
\begin{array}{ccccccccc}
0 & \longrightarrow & B_{i+1} & \longrightarrow & Z_i & \longrightarrow & H_i & \longrightarrow & 0 \\
& & \downarrow \varphi| & & \downarrow \varphi & & \downarrow & & \\
0 & \longrightarrow & \oplus \, n_i\Lambda & \longrightarrow & \oplus \Lambda & \longrightarrow & \oplus \Lambda/n_i\Lambda & \longrightarrow & 0.
\end{array}
$$

We can make φ onto by adding free summands to B_{i+1} and Z_i. Then, $\ker \varphi = \ker \varphi|$ is stably free. Hence, so is B_{i+1}.

Proposition I.2.4 (Universal Coefficient Theorem): If $\{C_*, \partial\}$ is a free Λ-chain complex with $C_i = 0$ for $i < 0$ and with $H_i(C_*)$ a nice Λ-module for all i, then there is a short exact sequence of Λ-modules.

$$0 \longrightarrow \mathrm{Hom}_\Lambda(\mathrm{Tor}\, H_{i-1}, \mathbb{Q}/\mathbb{Z} \otimes \Lambda) \longrightarrow H^i \longrightarrow \mathrm{Hom}_\Lambda(H_i/\mathrm{Tor}, \Lambda) \longrightarrow 0 \ .$$

Consequently, H^i is a nice Λ-module, and $\mathrm{Tor}\, H^i$ is isomorphic to $\mathrm{Hom}_\Lambda(\mathrm{Tor}\, H_{i-1}, \mathbb{Q}/\mathbb{Z} \otimes \Lambda)$.

Proof: The map $H^i \to \mathrm{Hom}_\Lambda(H_i, \Lambda)$ is given by evaluation of a cocycle representative on a cycle representative. The map $\mathrm{Hom}_\Lambda(\mathrm{Tor}\, H_{i-1}, \mathbb{Q}/\mathbb{Z} \otimes \Lambda) \to H^i$ is defined as follows. Let φ: $\mathrm{Tor}\, H_{i-1} \to \mathbb{Q}/\mathbb{Z} \otimes \Lambda$ is a Λ-homomorphism. Define $Z'_{i-1} \hookrightarrow Z_{i-1}$ to be the cycles of finite order in homology. Z'_{i-1} is stably free and in fact $Z_{i-1} \cong Z'_{i-1} \oplus H_{i-1}/\mathrm{Tor}$. Since Z'_{i-1} is projective, φ can be lifted to give a commutative diagram

$$0 \longrightarrow B_i \longrightarrow Z'_{i-1} \longrightarrow \operatorname{Tor} H_{i-1} \longrightarrow 0$$

$$\downarrow \bar{\varphi}| \qquad \downarrow \bar{\varphi} \qquad \downarrow \varphi$$

$$0 \longrightarrow \mathbb{Z} \otimes \Lambda \longrightarrow \mathbb{Q} \otimes \Lambda \longrightarrow \mathbb{Q}/\mathbb{Z} \otimes \Lambda \longrightarrow 0$$

The composition $C_i \overset{\partial}{\to} B_i \overset{\bar{\varphi}|}{\longrightarrow} \Lambda$ is a cocycle whose cohomology class is independent of the choice of $\bar{\varphi}$. The association $\varphi \to [\bar{\varphi}\partial]$ is a Λ-homomorphism $\operatorname{Hom}_\Lambda (\operatorname{Tor} H_{i-1}, \mathbb{Q}/\mathbb{Z} \otimes \Lambda) \overset{\lambda}{\to} H^i$.

The sequence

$$0 \longrightarrow \operatorname{Hom}_\Lambda (\operatorname{Tor} H_{i-1}, \mathbb{Q}/\mathbb{Z} \otimes \Lambda) \overset{\lambda}{\longrightarrow} H^i \overset{i}{\longrightarrow} \operatorname{Hom}_\Lambda (H_i/\operatorname{Tor}, \Lambda) \longrightarrow 0$$

is exact. The proof of this is the same as the proof of the usual universal coefficient theorem which uses only the fact that the modules of cycles and boundaries are projective.

__Theorem I.2.5:__ If $f: M^n \to N^n$ is a nice normal map, then

  1) $\operatorname{Ad}(\cdot): K_i(f)/\operatorname{Tor} \longrightarrow \operatorname{Hom}_\Lambda (K_{n-i}(f)/\operatorname{Tor}, \Lambda)$, and

  2) $\operatorname{Ad}(\ell): \operatorname{Tor} K_i(f) \longrightarrow \operatorname{Hom}_\Lambda (\operatorname{Tor} K_{n-i-1}(f), \Lambda)$ are isomorphisms.

(When the adjoint of either the intersection pairing or linking pairing is an isomorphism we say that the pairing is non-singular.)

__Proof:__ According to [15], p. 25, the chain complex for f, $C_*(f)$, satisfies Poincare duality: $\cap[M]: K^{n-i}(f) \overset{\cong}{\to} K_i(f)$. Thus we have

$$0 \longrightarrow \operatorname{Tor} K_i(f) \longrightarrow K_i(f) \longrightarrow K_i(f)/\operatorname{Tor} \longrightarrow 0$$

$$\downarrow \cong \qquad\qquad \downarrow \cong \qquad\qquad \downarrow \cong$$

$$0 \longrightarrow \operatorname{Hom}_\Lambda (\operatorname{Tor} K_{n-i-1}(f), \mathbb{Q}/\mathbb{Z} \otimes \Lambda) \longrightarrow K^{n-i}(f) \longrightarrow \operatorname{Hom}_\Lambda (K_{n-i}(f)/\operatorname{Tor}, \Lambda) \longrightarrow 0$$

Since Poincare duality between the simplicial chains for some triangulation of M and the cellular cochains on the dual cell decomposition is given by the intersection matrix between simplices and dual cells, the above Poincare duality isomorphisms are the same as adjoints to the linking and intersection pairings defined on these modules.

Corollary I.2.6: Suppose $i \neq n-i-1$. There are resolutions

$$0 \longrightarrow A_i \longrightarrow F_i \longrightarrow \text{Tor } K_i(f) \longrightarrow 0$$

and

$$0 \longrightarrow A_{n-i-1} \longrightarrow F_{n-i-1} \longrightarrow \text{Tor } K_{n-i-1}(f) \longrightarrow 0,$$

and chain realizations of them such that the maps induced by geometric
intersection of chains

$$I_1: F_i \times A_{n-i-1} \longrightarrow \Lambda \qquad \text{and} \qquad I_2: A_i \times F_{n-i-1} \longrightarrow \Lambda$$

are non-singular pairings.

Proof: Pick any isomorphism $\text{Tor } K_{n-i-1}(f) \cong \underset{j}{\oplus} \Lambda/n_j\Lambda$ and let
$0 \to A_{n-i-1} \to F_{n-i-1} \to \text{Tor } K_{n-i-1}(f) \to 0$ be the resolution corresponding
to it. There is a natural isomorphism of $\text{Hom}_\Lambda(\Lambda/n_j\Lambda, \mathbb{Q}/\mathbb{Z} \otimes \Lambda)$ with $\Lambda/n_j\Lambda$.
Thus we have

$$\text{Tor } K_i(f) \cong \text{Hom}_\Lambda(\text{Tor } K_{n-i-1}(f), \mathbb{Q}/\mathbb{Z} \otimes \Lambda) \cong \text{Hom}(\oplus \Lambda_j/n_j\Lambda_j, \mathbb{Q}/\mathbb{Z} \otimes \Lambda) \cong \oplus \Lambda_j/n_j\Lambda_j.$$

Use this isomorphism to induce a resolution for $\text{Tor } K_i(f)$. An algebraic
map lifting the linking pairing, $I: F_i \times F_{n-i-1} \to \mathbb{Q} \otimes \Lambda$ can be taken to
be the diagonal matrix under the natural bases

$$\begin{pmatrix} \frac{1}{n_i} \otimes e & & 0 \\ & \ddots & \\ 0 & & \frac{1}{n_r} \otimes e \end{pmatrix}$$

The induced pairings I_1 and I_2 are both given by the identity matrix.
Lemma 2.2 implies that I comes from some chain realizations of the
resolutions.

To calculate surgery obstructions in $L_n^s(\pi)$, we must work with based

Λ-chain complexes instead of just homology groups. Because of this, we introduce the notion of a nice Λ-chain complex.

Definition: A nice Λ-chain complex is one which is isomorphic to a direct sum of finitely many complexes of the form $\{0 \to \Lambda \to 0\}$ and $\{0 \to \Lambda \xrightarrow{\times n} \Lambda \to 0\}$ for $n \in \mathbb{Z}^+$.

The homology of a nice Λ-chain complex is a nice Λ-module. In fact, there are canonical isomorphisms $\mathrm{Tor}\, H_i \cong \oplus \Lambda/n_j\Lambda$ and $H_i/\mathrm{Tor} \cong \oplus \Lambda$ for any nice Λ-complex.

Lemma I.2.7: Let $\{C_*, \partial\}$ be a free Λ-chain complex with only finitely many non zero homology modules each of which is nice. For any collections of isomorphisms $\mathrm{Tor}\, H_i(C_*) \cong \oplus \Lambda/n_j\Lambda$ and $H_i(C_*)/\mathrm{Tor} \cong \oplus \Lambda$, there is a homotopy equivalence of a nice Λ-complex with C_* inducing these isomorphisms. If C_* is a based Λ-complex, then all maps of nice Λ-complexes   realizing a fixed set of isomorphisms have the same Whitehead torsion.

Proof: The first half of the lemma is straightforward. Given two such maps realizing the same set of isomorphisms we will find a sequence of maps connecting them. Each term in the sequence will differ from its predecessor either by chain homotopy or by adding multiples of one basis element to another. Thus, all maps in the sequence will have the same Whitehead torsion. First on the generators corresponding to $\mathrm{Tor}\, H_i$, since the maps are the same on homology there is a chain homotopy connecting them. On the generators corresponding to H_i/Tor, the maps differ by a torsion element and a homology. Thus by adding multiples of the torsion generators to the free generators and performing another chain homotopy we can make the maps agree here. Lastly on the chains which give the relations in $\mathrm{Tor}\, H_i$, their boundaries are the same, and hence their differences are $(i+1)$ cycles. By adding multiples of the $(i+1)$ torsion

and free generators and performing a chain homotopy we can make them
agree here also.

Definition I.2.8: Let (C_*, ∂) be a based, free Λ-chain complex as above.
A basis for $H_*(C_*)/\text{Tor}$ and an isomorphism $\text{Tor } H_*(C_*) \cong \oplus \Lambda/n_j\Lambda$ is called
a <u>based structure for</u> $H_*(C_*)$ if any (and therefore all) maps of a nice
Λ-complex onto C_* realizing these isomorphisms are simple homotopy
equivalences.

Definition I.2.9: Let $f: M^n \to N^n$ be a nice normal map which is a simple
homotopy equivalence on ∂M. Then, $C_*(f)$ has a simple equivalence class
of bases, and hence $K_*(f)$ has a based structure. We say that f is
<u>s-nice</u> if it has a based structure so that

1) the intersection pairings have adjoints which are simple isomorphisms, and

2) the linking pairings lift to intersection maps on the resolutions,
   $I: F_i \times F_{n-i-1} \to \mathbb{Q} \otimes \Lambda$ so that $\text{ad}(I): A_i \to F^*_{n-i-1}$ is a simple
   isomorphism for $i \neq n-i-1$.

Note: 1) For any nice normal map, 2.6 implies that there is a nice
   Λ-complex mapping in by a homotopy equivalence so that 1) and 2)
   are satisfied.

2) The argument in 2.7 shows that it is always possible to assume
   the chains giving the relations in $\text{Tor } K_{n-i-1}$ are disjoint from
   the i-cycles.

Section I.3 - Low dimensional surgery. Let $f: M^n \to N^n$ be an s-nice normal
map. We prove that we can do surgery to produce a normal bordism from
f to a highly connected normal map while keeping track of the kernel
modules and their based structure. The results of this section are
accumulated in the following theorem.

Theorem I.3.4: If $f: M^n \to N^n$ is an s-nice normal map, then f is normally bordant to $f': M' \to N$ such that:

1) if $n = 2s + 1$, then $K_i(f') = 0$ for $i \neq s$, $K_s(f') = \text{Tor } K_s(f)$ as a module with based structure, and the linking pairing on $K_s(f')$ equals that on $\text{Tor } K_s(f)$;

2) if $n = 2s$, then $K_i(f') = 0$ for $i \neq s$, $K_s(f') = K_s(f)/\text{Tor} \oplus \Lambda^r \oplus \Lambda^r$ as based Λ-modules, and the intersection pairing on $K_s(f')$ is given by

$$\begin{pmatrix} \lambda_f & 0 & * \\ \hline 0 & 0 & (-1)^s \text{Id} \\ \hline * & \text{Id} & * \end{pmatrix}$$

with λ_f the original intersection pairing on $K_s(f)/\text{Tor}$.

3) If f is nice instead of s-nice, then 1) and 2) above are true if the conditions on the based structure are omitted.

Wall proves in general that it is always possible to "concentrate" the kernel groups in the middle dimensions. The extra result here is the possibility of keeping track of the modules and pairings. This will lead us to a priori definitions of the surgery obstruction for an s-nice or nice normal map. All the proofs will be given for an s-nice normal map. Each step is valid for a nice normal map if all references to the based structure is omitted.

Proposition I.3.1: Let $f: M^n \to N^n$ be an i-connected, s-nice, normal map, $i < [\frac{n}{2}]$. We can do surgery to produce a normal bordism $G: W^{n+1} \to N^n \times I$ from f to $f': M' \to N$ such that f' is a s-nice normal map and

$$K_*(f') = \begin{cases} K_*(f) & * \neq i, \, n - i \\ \text{Tor } K_i(f) & * = i \\ 0 & * = n - i. \end{cases}$$

As always, the equal sign means as modules with based structure.

<u>Proof</u>: Let $\{x_1,\ldots,x_r\}$ be the elements in $K_i(f)$ which form the basis for
$K_i(f)/\mathrm{Tor}$. We realize the x_i by disjointly embedded spheres, $\{S_1^i,\ldots,S_r^i\}$,
by general position. Using the fact that $K_i(f) = \pi_{i+1}(N,M)$ these spheres
bound disks $D_j^{i+1} \to N$. We use these disks to give a trivialization of
$\nu_{S^i} - \nu_M|S^i$. (ν_X means the stable normal bundle of the X.) Such
trivializations give the embedded spheres trivialized normal bundles.
Let $G: W^{n+1} \to N \times I$ be the trace of surgery along these spheres. That is
W is $M \times I \cup (D_1^{i+1} \times D^{n-i} \cup \ldots \cup D_r^{i+1} \times D^{n-1})$ where the handles
$\{D_j^{i+1} \times D^{n-i}\}$ are added along the spheres.

Denote the cores of the handles union the spheres cross I by
$\{d_1,\ldots,d_r\}$ and let d_i' be the dual (n-i)-disk to d_i.

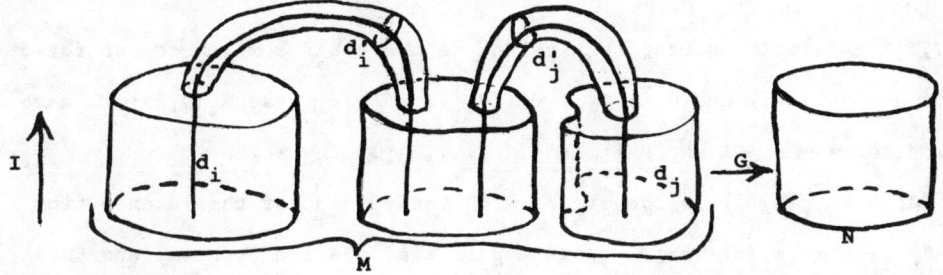

We have the short exact sequence of chain groups with 0 Whitehead
torsion

(*) $0 \longrightarrow C_*(f) \longrightarrow C_*(G) \longrightarrow C_*(G,f) \longrightarrow 0,$

and the resulting long exact sequence of kernel groups.

$$\cdots \longrightarrow K_{*+1}(G,f) \xrightarrow{\partial} K_*(f) \longrightarrow K_*(G) \longrightarrow \cdots .$$

As a based chain complex $C_*(G,f)$ is simple homotopy equivalent to the
chain complex with $C_{i+1} = \Lambda(d_1,\ldots,d_r)$ and all other modules equal zero.
This, then, is also $K_*(G,f)$.

Since $\partial(d_i) = x_i$, we see from the long exact sequence above that

$$K_*(G) = \begin{cases} K_*(f) & * \neq i, \\ \text{Tor } K_i(f) & * = i. \end{cases}$$

Let $C_* \to C_*(f)$ be a nice Λ-complex associated to the based structure.
From sequence (*) we see that a nice Λ-complex for $C_*(G)$ is obtained by
deleting the free summands corresponding to $K_i(f)/\text{Tor}$ from C_*. Thus, the
above isomorphism is as modules with based structure.

We also have the exact sequence for the pair (W, M'):

$$0 \longrightarrow C_*(f') \longrightarrow C_*(G) \longrightarrow C_*(G, f') \longrightarrow 0,$$

and the resulting long exact sequence

$$\cdots \longrightarrow K_*(f') \longrightarrow K_*(G) \longrightarrow K_*(g, f') \longrightarrow \cdots .$$

$C_*(G, f')$ is simple homotopy equivalent to the chain complex concentrated
in dimension $(n-i)$ and with $C_{n-i} = \Lambda(d_1', \ldots, d_r')$. Thus, $K_*(G, f')$ is zero
except for $* = n - i$ where it is $\Lambda(d_1', \ldots, d_r')$. The map
$K_{n-i}(G) \to K_{n-i}(G, f')$ is identified with the adjoint of the intersection
map $K_{n-i}(f) \to [K_i(f)/\text{Tor}]^*$. This map is a simple isomorphism, and thus
$K_{n-i}(f') = 0$. The chains of dimension $(n-i)$ in the resolution for
$\text{Tor } K_{n-i-1}(f)$ can be assumed disjoint from the spheres on which we did
surgery. Thus, they persist to M' to give a chain realization for
$\text{Tor } K_{n-i-1}(f')$. (This is automatically true for the chain realizations
for $\text{Tor } K_*(f) * < n - i - 1$.) Thus

$$K_*(f') = \begin{cases} K_*(f) & * \neq i, n - i \\ \text{Tor } K_i(f) & * = i \\ 0 & * = n - i \end{cases}$$

as modules with bases structure. All the intersection pairings and linking pairings for $K_*(f')$ agree with these for $K_*(f)$. Consequently, f' is
a nice normal map.

We turn now to the case of surgery to kill a torsion subgroup.

Proposition I.3.2: Let $f: M^n \to N^n$ be an s-nice, normal map which is i-connected for $i < [\frac{n-1}{2}]$ and with $K_i(f) = \text{Tor } K_i(f)$. We can perform surgery on f to produce a normal bordism from f to f' with f' an s-nice normal map and

$$K_*(f') = \begin{cases} K_*(f) & \text{for } * \neq i, i+1, n-i-1 \\ 0 & \text{for } * = i \\ K_i(f) \oplus \Lambda^r & \text{for } * = i+1 \\ K_{n-i-1}(f)/\text{Tor} \oplus \Lambda^r & \text{for } * = n-i-1, \end{cases}$$

as modules with based structure.

Proof: There are resolutions for $\text{Tor } K_i(f)$ and $\text{Tor } K_{n-i-1}(f)$

$$0 \longrightarrow A_i \overset{\varphi_i}{\longrightarrow} F_i \longrightarrow \text{Tor } K_i(f) \longrightarrow 0,$$

and

$$0 \longrightarrow A_{n-i-1} \overset{\varphi_{n-i-1}}{\longrightarrow} F_{n-i-1} \longrightarrow \text{Tor } K_{n-i-1}(f) \longrightarrow 0$$

which have chain realizations with chain intersections inducing simple isomorphisms

$$I_i: A_{n-i-1} \longrightarrow F_i^* \quad \text{and} \quad I_{n-i-1}: A_i \longrightarrow (F_{n-i-1})^*.$$

We can assume the basis for F_i is realized by disjointly embedded spheres with trivialized normal bundles $\{S_1^i, \ldots, S_r^i\} \to M'$, as before. Let the basis for A_i be realized by chains $\{C_1^{i+1}, \ldots, C_r^{i+1}\}$ with $\partial C_j = n_j S_j^i$. Let the chain realization for the resolution of $\text{Tor } K_{n-i-1}(f)$ be chains $\{Y_j^{n-i}\}$ and cycles $\{Z_j^{n-i-1}\}$ with $\partial Y_j = n_j Z_j$.

Let $G: W^{n+1} \to N^n \times I$ be the trace of surgery on these spheres, with $f': M' \to N$ the result of the surgery. Let the handles added, union the spheres cross I in $M \times I$, be $\{d_1, \ldots, d_r\}$ with their dual handles $\{d_1', \ldots, d_r'\}$.

From the short exact sequence for the pair (G, f),

$$0 \longrightarrow C_*(f) \longrightarrow C_*(G) \longrightarrow C_*(G,f) \longrightarrow 0$$

and the associated long exact sequence

$$\cdots \longrightarrow K_*(f) \longrightarrow K_*(G) \longrightarrow K_*(G,f) \overset{\partial}{\longrightarrow} K_{*-1}(f) \longrightarrow \cdots$$

we see that

$$K_*(G) = \begin{cases} K_*(f) & * \neq i,\ i+1 \\ 0 & * = i \\ K_{i+1}(f) \oplus A_i & * = i+1 \end{cases}$$

as modules with bases structures.

The splitting of

$$0 \longrightarrow K_{i+1}(f) \longrightarrow K_{i+1}(G) \longrightarrow A_i \longrightarrow 0$$

is given by the chain realization for A_i. Namely, the basis element $a_j \in A_i$ goes to cycle $(- C_j \cup n_j d_j)$ in W.

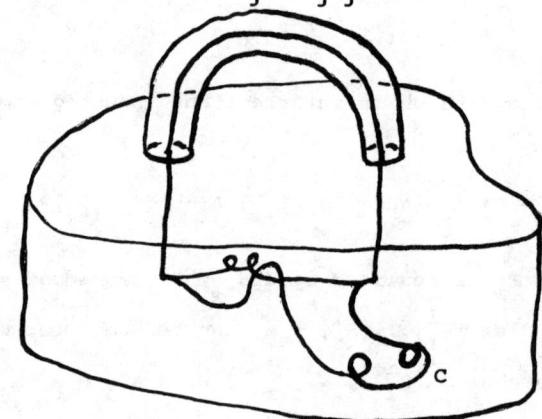

The long exact sequence for the pair (G,f') is

$$\cdots \longrightarrow K_{n-i}(F) \longrightarrow K_{n-i}(G,f') \longrightarrow K_{n-i-1}(f') \longrightarrow K_{n-i-1}(G) \longrightarrow 0 \longrightarrow \cdots$$

$$\Big\downarrow = \qquad\qquad \Big\downarrow = \qquad\qquad\qquad\qquad \Big\downarrow =$$

$$0 \qquad\qquad \wedge(d_1', \ldots, d_r') \qquad\qquad\qquad\qquad K_{n-i-1}(f)$$

Since the cycle representatives which give a basic for $K_{n-i-1}(f)/\text{Tor}$ and also the $\{z_j^{n-i-1}\}$ lie disjoint from the spheres on which we did surgery, they persist to form cycles in $K_{n-i-1}(f')$. This gives a map

$$K_{n-i-1}(f)/\mathrm{Tor} \oplus F_{n-i-1} \longrightarrow K_{n-i-1}(f').$$

<u>Claim</u>: The map $\Lambda(d_1',\ldots,d_r') \to K_{n-i-1}(f')$ has image contained in the image of F_{n-i-1}. The map is

$$F_i^* \xrightarrow{\mathrm{ad}(I)^{-1}} A_{n-i-1} \hookrightarrow F_{n-i-1} \hookrightarrow K_{n-i-1}(f').$$

<u>Corollary</u>: $K_{n-i-1}(f')/\mathrm{Tor} \oplus F_{n-i-1} \to K_{n-i-1}(f')$ is an isomorphism.

<u>Proof of Claim</u>: The chain intersection map between the $\{Y_j\}$ and the spheres on which we do surgery defines $\mathrm{ad}(I): A_{n-i-1} \to F_i^*$. If we cut out disks around the intersections of the Y_j with the spheres, then the Y_j provides a homology in M' from $n_j z_j$ to $\mathrm{ad}(I)$ $([Y_j])$.

We have now shown :

$$K_*(f') = \begin{cases} K_*(f) & * \neq i, i+1, n-i-1 \\ 0 & * = i \\ K_{i+1}(f) \oplus A_i & * = i+1 \\ K_{n-i-1}(f)/\mathrm{Tor} \oplus F_{n-i-1} & * = n-i-1 \end{cases}$$

These isomorphisms are all obviously isomorphisms of modules with based structure except in dimension $n - i - 1$. Since $\mathrm{ad}(I_{n-i-1})^{-1}: F_i^* \to A_{n-i-1}$ is a simple isomorphism, it is also true in this dimension. Under the splittings given above $\lambda_{f'}: K_{i+1}(f') \times K_{n-i-1}(f') \to \Lambda$ is given by the matrix

$$\begin{pmatrix} \lambda_f & 0 \\ 0 & -I_{n-i-1} \end{pmatrix} ,$$

where λ_f is the original pairing for f. This is proved by looking at the cycle representatives we have given for the various classes. All other intersection and linking pairings are unchanged (except of course for ℓ: Tor $K_i(f) \times$ Tor $K_{n-i-1}(f) \to \mathbb{Q}/\mathbb{Z} \otimes \Lambda)$. Thus f' is an s-nice normal map.

This argument also calculates the effect of surgery to kill Tor $K_{n-1}(f)$ for f: $M^{2n} \to N^{2n}$. Since the answer is a little different, we state it as a separate proposition.

<u>Proposition</u> I.3.3: Let f: $M^{2n} \to N^{2n}$ be an s-nice, normal map which is (n-1) connected and with Tor $K_{n-1}(f) = K_{n-1}(f)$. We can do surgery to produce a normal bordism for f to f' such that f' is s-nice and

$$K_*(f') = \begin{cases} 0 & * \neq n \\ K_n(f)/\text{Tor} \oplus \Lambda^r \oplus \Lambda^r & * = n \end{cases}$$

as based Λ-modules. The intersection pairing on $K_n(f')$ is given by the matrix

$$\begin{pmatrix} \lambda_f & 0 & * \\ 0 & 0 & -I_n \\ * & (-1)^{n+1}\bar{I}_n & * \end{pmatrix}.$$

<u>Proof</u>: This is proved by the argument used in 3.2. The difference is that both A_{n-1} and F_n are added to $K_n(f')$. The cycle representatives allow us to calculate the matrix of intersections.

Note that the first Λ^r factor, the one corresponding to F_n, is generated by cycles lying geometrically in M. In M, they are torsion cycles. Thus, they have zero intersection with any class in M' which is homologous in W to a class lying in M.

Summing up, we have shown the following.

<u>Theorem</u> I.3.4: If f: $M^n \to N^n$ is an s-nice, normal map, then we can perform surgery to produce a normal bordism from f to f': $M'^n \to N^n$ such that

a)   if $n = 2s + 1$, then $K_i(f') = 0$ for $i \neq s$, $K_s(f') = \text{Tor } K_s(f)$ as
     modules with bases structure, and the linking pairing on $K_s(f')$
     equals that on $\text{Tor } K_s(f)$; and

b)   if $n = 2s$, then $K_i(f') = 0$ for $i \neq s$, $K_s(f') =$
     $K_s(f)/\text{Tor} \oplus \Lambda^r \oplus \Lambda^r$ as based modules, and the intersection
     pairing on $K_s(f')$ is equal to

$$\begin{pmatrix} \lambda_f & 0 & * \\ 0 & 0 & -I_n \\ * & (-1)^{n+1}\bar{I}_n & * \end{pmatrix}$$

with I_n a simple isomorphism.

c)   If $f: M^n \to N^n$ is a nice, normal map, then a) and b) are true
     after omitting the reference to based structures.

CHAPTER II: An A-priori Definition of the Surgery Obstruction

Section II.1 - Case I - The even dimensions. Now we turn to the problem of calculating the surgery obstruction of a nice, or s-nice, normal map before we actually do the surgery to make it highly connected. We will find (G,λ,μ) satisfying the properties to define an element in $L_{2n}^{s}(\pi)$ associated to an s-nice, normal map $f: M^{2n} \to N^{2n}$. This triple will be geometrically defined without assuming that f is highly connected. By doing surgery, we show that this triple determines the usual Wall surgery obstruction, $\sigma(f) \in L_{2n}^{s}(\pi)$. Deleting the parts of the discussion dealing with the based structure, produces a triple (G,λ,μ) associated to a nice normal map and proves that it gives the Wall surgery in $L_{2n}(\pi)$. Let $f: M^{2n} \to N^{2n}$ be an s-nice surgery problem, covered by the bundle map $\tilde{f}: \nu_M \to \zeta$.

Our first guess for the free Λ module G is $K_n(f)/\text{Tor}$. It is equipped with a basis and already has a pairing $\lambda: K_n(f)/\text{Tor} \times K_n(f)/\text{Tor} \to \Lambda$ which satisfies

1) λ is Λ linear in the second variable,

2) λ is non-singular, and

3) $\lambda(x,y) = (-1)^n \overline{\lambda(y,x)}$.

$\text{Ad}(\lambda)$ is a simple isomorphism. To enhance $(K_n(f)/\text{Tor},\lambda)$ so that it defines an element in $L_{2n}^{s}(\pi)$ a "μ form" is required, $\mu: K_n(f)/\text{Tor} \to Q_n = \Lambda/\{\nu - (-1)^n \bar{\nu}\}$. The μ-form comes from the geometric self-intersection number for any element in $K_n(f)$, using the Rourke-Sullivan idea of immersed cycles. For the prototype of this argument see [11].

If $V^n \underset{i}{\rightarrowtail} M^{2n}$ is a based immersion of a simply connected manifold (i.e. V actually immerses in $\tilde{M}^{2n}$ and the projects down), then there is a self-intersection number for this immersion, as described in section 1.

It is an invariant of the regular homotopy class of the immersion.

The self intersection $\mu_f \colon K_n(f) \to Q_n$ is defined by using the bundle map $\tilde{f}$ to pick out a regular homotopy class of immersed submanifolds representing $x \in K_n(f)$. First, we note that it is sufficient to define $\mu_f(rx)$ for r odd. This follows because

1)   as an abelian group Q_n has no odd torsion, and

2) $\mu_f(rx) = r^2 \mu_f(x)$ for $r \in \mathbb{Z}$.

For some odd integer r, $rx \in K_n(f) = H_{n+1}(\tilde{N}, \tilde{M})$ is represented by a relative bordism element

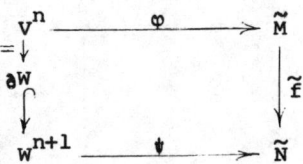

see [3].

Given such a (V, W, φ, ψ) representing rx, then the bundle $\nu_W - \psi^*(\xi)$ reduces to an n-dimensional bundle over W^{n+1} (since W is homotopy equivalent to an n-complex). Any such reduction induces by restriction a reduction of $\nu_V - \varphi^* \nu_{\tilde{M}}$. But an n-dimensional reduction of this bundle is equivalent to an immersion of $V^n \looparrowright \tilde{M}$, [5], homotopic to φ. Define $\mu_f(rx)$ to be the self-intersection of $V^n \looparrowright M$ for any immersion obtained from a reduction of $\nu_W - \psi^*(\xi)$.

Proposition II.1.1: $\mu_f(rx)$ is well defined independent of all the choices above. $\mu_f(rx)$ is divisible by r^2 in Q_n (and thus uniquely divisible by r^2 since r is odd). Define $\mu_f(x) = \frac{1}{r^2} \mu_f(rx)$. Then $\mu_f \colon K_n(f) \to Q_n$, satisfies

1) $\mu_f(x \cdot a) = \bar{a} \mu(x) a$ for $a \in \Lambda$,

2) $\mu_f(x+y) = \mu_f(x) + \mu_f(y) + \lambda(x,y)$ in Q_n, and

3) $\mu_f(x) + (-1)^n \mu_f(\bar{x}) = \lambda(x,x)$ in Λ.

Proof: By I.3.4 we can do surgery on f to make $K_i(f) = 0$ for $i \leq n - 2$.

This does not change $K_n(f)$ or the intersection form on $K_n(f)/\text{Tor}$. If $r \cdot x$ is represented by (V,W,φ,ψ), we can do the low dimensional surgery away from the image of $V^n \to M^{2n}$. All geometric information required to define $\mu(rx)$ is unchanged, and $\mu(rx)$ calculated before surgery equals $\mu(rx)$ calculated after surgery. Thus, it suffices to prove II.1.1 for normal maps which are $(n-2)$ connected. Since $\Omega_*(X,Y) = H_*(X,Y)$ for $* \leq$ (connectivity of $(X,Y) + 3$), for a $(n-2)$ connected normal map all $x \in K_n(f)$ are represented by (V,W,φ,ψ) as above and any two representatives (V,W,φ,ψ) and (V',W',φ',ψ') are bordant by some

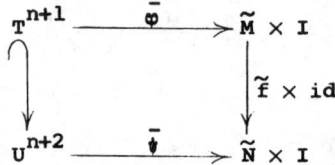

Let ζ^n be a reduction of $\nu_V - \varphi^*(\nu_{\widetilde{M}})$ induced by the first representative. Extend this to a reduction of $\nu_T - \bar{\varphi}^*(\nu_{\widetilde{M}})$ to an n-dimensional bundle $\bar{\zeta}^n$. This is possible since $H_*(T,V) = 0$ for $* > n$. $\bar{\zeta}^n|V'$ gives a reduction of $\nu_{V'} - \varphi'^*(\nu_{\widetilde{M}})$ to $\bar{\zeta}'^n$.

<u>Claim</u> II.1.2: If we use ζ^n to immerse V in M and $\bar{\zeta}'^n$ to immerse V' in M, then the self-intersections of V and V' in M agree.

<u>Proof</u>: $\bar{\zeta}^n$ gives an immersion of T into $M \times I$ connecting these two immersions of V and V'. $T \cdot T$ is a 1-manifold whose boundary, on the one hand, is 0 and, on the other, is (self-intersection of V) - (self-intersection of V').

<u>Claim</u> II.1.3: Using $\bar{\zeta}'^n$ as above to immerse V' gives the same self-intersection as using any reduction induced from a reduction of $\nu_{W'} - \psi'^*(\zeta)$.

<u>Proof</u>: (See, for example, [11], and [9] chapter 5.)

We have $V' \xrightarrow{\varphi'} \widetilde{M}$ bounding $W' \xrightarrow{\psi'} \widetilde{N}$ and $W'' \xrightarrow{\psi''} \widetilde{N}$, $(W'' = TU - W)$ and two

immersions of $V' \looparrowright M$ induced from bundle reductions of $\nu_{W'} - \psi'^*(\xi)$ and
$\nu_{W''} - \psi''^*(\xi)$, respectively. The induced reductions over V' clearly agree
as $(n+1)$ reductions since V' is an n-complex. Thus together they define
a reduction of $\nu_{W' \cup W''} - (\psi' \cup \psi'')^*(\xi)$ to an $(n+1)$ bundle η^{n+1}

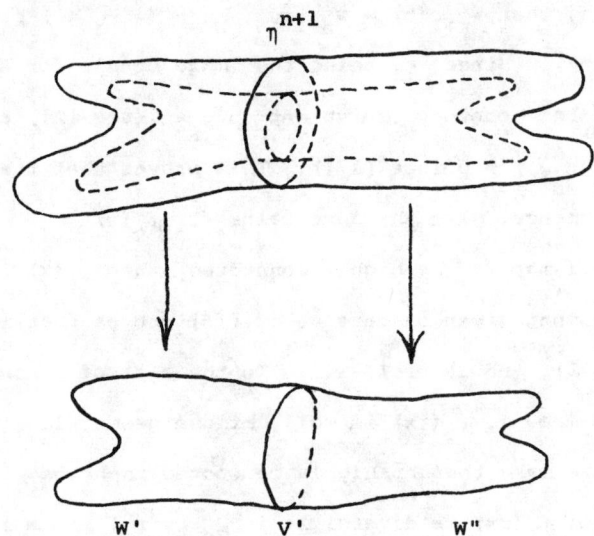

The difference of the two bundle reductions over V'^n is an element in
$[V^n, F^n]$ where F^n is the fiber of $BSO(n) \to BSO$. $\pi_i(F^n) = 0$ for $i < n$ and

$$\pi_n(F^n) = \begin{cases} \mathbb{Z} & \text{n is even} \\ \mathbb{Z}/2 & \text{n is odd.} \end{cases}$$

Thus

$$[V^n, F^n] = \begin{cases} \mathbb{Z} & \text{n is even} \\ \mathbb{Z}/2 & \text{n is odd.} \end{cases}$$

It is easy to identify the difference of the two n-dimensional reductions
as the obstruction to a section (a section mod 2 if n is odd) of the
$(n+1)$-dimension bundle η^{n+1} over W. The bundle η^{n+1} is an oriented
bundle over an oriented $(n+1)$-manifold. If $(n+1)$ is odd, it has a section.
Thus, the two reductions agree in this case. If $(n+1)$ is even, then the

difference of two reductions as an element in $\mathbb{Z}/2$ is $w_{n+1}(\eta^{n+1})$. A
result of Atiyah, [1], says that the Thom space of the bundle $(\nu_L - \gamma)$
over a closed, oriented manifold L is Spanier-Whitehead dual to the
Thom space of γ over L. This implies that $w_{n+1}(\nu_L - \gamma) = v_{n+1}(\gamma)$, see
[9]. In our case, then $w_{n+1}(\eta) = w_{n+1}(\nu_{W'\cup W''} - (\psi' \cup \psi'')^*\xi)$
$= v_{n+1}((\psi' \cup \psi'')^*\xi)$. Since ξ, being the image bundle for a degree one
normal map, is fiber homotopy equivalent to $\nu_{N^{2n}}$ (see [2]) the Wu relations imply $v_{n+1}(\nu_{N^{2n}}) = 0$ (see [17]). This proves that the two reductions agree and, hence, give the same value for $\mu_g(x)$.

If the normal map f is highly connected, then $\mu_f(x)$ is defined for
all x. The argument given on page 46 of [15] proves that in this case
it satisfies 1), 2), and 3) of II.1.1. In the case of a non-highly
connected normal map f, $\mu_f(rx)$ is well defined geometrically for some r.
In addition, if we make the map highly connected then the μ-form is
unchanged. Hence, $\mu_f(rx)$ is diversible in Q_n by r^2, since it is unchanged
by surgery and is equal to $r^2\mu(x)$ after surgery. If we define
$\mu_f(x) = \dfrac{1}{r^2}\mu_f(rx)$ in the non-highly connected case, then this μ-form
agrees with the geometrically defined one after we make the map highly
connected. Consequently it satisfies II.1.1,1), 2), and 3).

Note: In the end we have a definition of μ_f before doing surgery. We
only need to do surgery to show that it is well defined and satisfies 1),
2), and 3).

If we partition the elements of $\pi = \pi_1(N)$ into $\{g_i, g_i^{-1}\}$, $\{t_i\}$, $\{s_i\}$
where $g_i \neq g_i^{-1}$, $t_i = t_i^{-1}$ and $w(t_i) = (-1)^n$, and $s_i = s_i^{-1}$ and
$w(s_i) = (-1)^{n+1}$, then $Q_n \cong \underset{\{g_i\}}{\oplus} \mathbb{Z} \oplus \underset{\{t_i\}}{\oplus} \mathbb{Z} \oplus \underset{\{s_i\}}{\oplus} \mathbb{Z}/2\mathbb{Z}$.

$$\lambda(x,x) = \sum_{g_i}\left[\lambda_{g_i}\cdot g_i + (-1)^n w(g_i)\lambda_{g_i}\cdot g_i^{-1}\right] + \sum_{t_i} 2\lambda_{t_i}\cdot t_i,$$

and

$$\mu(x) = \sum_{g_i} \lambda_{g_i} \cdot g_i + \sum_{t_i} \lambda_{t_i} \cdot t_i + \sum_{s_i} \epsilon_i \cdot s_i$$

where $\epsilon_i = 0$ or 1. Thus, the only new algebraic information contained in $\mu(x)$ is the $\mathbb{Z}/2$ coefficients of those elements which are their own inverses and which act by $(-1)^{n+1}$ on the orientation of N. All the coefficients, ϵ_i, except the coefficient of the identity element are determined by intersection data in the universal cover. The coefficient of the identity is the only one that requires the immersion idea.

Now we wish to use II.1.1 to define $\mu_f\colon K_n(f)/\text{Tor} \to Q_n$ which, together with the intersection pairing, will define an element in $L_n^s(\pi)$. Unfortunately, μ_f may not vanish on Tor $K_n(f)$. In fact, there are simply connected examples where it does not. However, if $\mu_f|\text{Tor } K_n(f)$ is 0, then by II.1.1, it does indeed define a map $\mu_f\colon K_n(F)/\text{Tor} \to Q_n$.

For the kernel groups encountered in proving the product formula, μ_f will vanish on the torsion. We make the following

   Assumption: μ_f vanishes on Tor $K_n(f)$.

Thus we now have an a priori surgery obstruction assigned to f, $\sigma(f) \in L_{2n}^s(\pi)$.

Theorem II.1.4: a) If $f\colon m^{2n} \to N^{2n}$ is an s-nice normal map, then it is normally cobordant to $f'\colon M' \to N$ such that f' is s-nice and

   1) $K_*(f') = 0$ for $* \neq n$,

   2) $K_n(f') = K_n(f)/\text{Tor}$ as based modules, and

   3) the geometrically defined λ and μ forms for f' and f agree.

If f is a nice normal map, then the above is true after ignoring the bases.

Proof: We use I.3.4 to replace $f\colon M^{2n} \to N^{2n}$ by $f'\colon M' \to N$ such that

   1) f' is s-nice,

---

2) $K_i(f') = 0$ for $i \neq n$, and

3) $K_n(f') = K_n(f)/\mathrm{Tor} \oplus \Lambda^r \oplus \Lambda^r$ as based Λ-modules.

The first Λ^r factor in $K_n(f')$ is generated by cycles lying geometrically in M and representing torsion classes in $H_*(M;\Lambda)$. Since the μ-form is geometrically defined, μ_f and $\mu_{f'}$ take the same value on these cycles. By our assumption, μ_f vanishes on them. Hence, so does $\mu_{f'}$.

The intersection form on $K_n(f')$ is given by the matrix

$$\begin{pmatrix} \lambda_f & 0 & * \\ 0 & 0 & -I_n \\ * & (-1)^{n+1}\bar{I}_n & * \end{pmatrix}$$

with I_n a simple isomorphism. According to [15], theorem 5.2, spheres representing a basis for the first Λ^r-factor can be disjointly embedded with trivial normal bundles. The same analysis as in I.3.1 shows that the result of performing surgery on these classes is to produce an s-nice normal map $f'' : M'' \to N$ with

1) $K_i(f'') = 0$ for $i \neq n$

2) $K_n(f'') = K_n(f)/\mathrm{Tor}$ as based modules,

3) the intersection forms on $K_n(f'')$ and $K_n(f)/\mathrm{Tor}$ are equal, and

4) $\mu_{f''} = \mu_f$.

Note: We have a choice for the cycles representing the second Λ^r factor. They are of the form $A_i \cup n_i d_i$ where $\{A_1^n,\ldots,A_r^n\}$ are chains giving a basis for A_{n-1} in the resolution $0 \to A_{n-1} \to F_{n-1} \to \mathrm{Tor}\, K_{n-1}(f) \to 0$. By choosing the A_i correctly, we can make this factor a subkernel of the $\Lambda^r \oplus \Lambda^r$ also. Then we could do surgery on it. Low dimensional surgery followed by this would give a normal bordism, W, from f to f'' as above with $K_n(W,f) \xrightarrow{\ } K_{n-1}(f)$ an isomorphism.

Theorem II.1.5: If $f: M^{2n} \to N^{2n}$ is an s-nice normal map, then the triple

$\{K_n(f)/\text{Tor},\lambda,\mu_f\}$ determines the Wall surgery obstruction of f in

$L^s_{2n}(\pi)$. If f: $M^{2n} \to N^{2n}$ is a nice normal map, then the triple

$\{K_n(f)/\text{Tor},\lambda,\mu\}$, where $K_n(f)/\text{Tor}$ has no distinguished basis, determines

the surgery obstruction of f in $L_{2n}(\pi)$.

Proof: Assume that $K_i(f) = 0$ for i $\neq$ n, and that $K_n(f)$ is free. Our λ

pairing and Wall's agree by definition.

  Wall defines his μ-form by using immersed spheres and taking their

self-intersections. The immersions of the spheres come from a triviali-

zation of their stable normal bundles. These trivialization are provided

by the fact that the spheres bound disks, $D^{n+1} \to N^{2n}$. This procedure is

just an example of our general procedure and hence defines the same

μ-form. Wall defines the basis for $K_n(f")$ by using the bases chain

complex $C_*(f")$. Our definition of the basis of $K_r(f")$ is that is comes

from the bases of a nice Λ-chain complex simple homotopy equivalent to

$C_*(f)$. Hence, the bases are simple equivalent. This shows that for (n-1)

connected normal maps Wall's triple and our triple agree. Applying

II.1.4 gives a proof of the theorem.

Note: If μ_f does not vanish on Tor $K_n(f)$, then it is still possible to

give an a-priori description of the surgery obstruction. One uses

  $K_n(f)$, λ, and μ on all of $K_n(f)$ to produce a triple which defines

$\sigma(f)$ in $L^s_{2n}(\pi)$ or in $L_{2n}(\pi)$.

Section II.2 - Case II - The odd dimensions. In this section we give

an a priori definition of the surgery obstruction of an (s-) nice normal

map between odd dimensional manifolds. Just as the case of even dimen-

sional manifolds is a generalization of the work of [11] on the Kervaire

invariant, the odd dimensional case is a generalization of the work of

sections 5 and 6 of [9] on odd dimensional normal maps between

$\mathbb{Z}/n$-manifolds.

The form μ_f used in the even dimensions is replaced in this case
by a quadratic refinement, q_f, of the middle dimensional linking pairing.
It is defined similarly to the μ-form. The extra information again comes
from the bundle map. This time it produces classes of $(n-1)$ dimensional
reductions of the stable normal bundles of $(n-1)$ manifolds mapping into
the domain, M^{2n-1}. Such reductions are equivalent to embeddings of the
$(n-1)$-manifolds together with nowhere zero normal fields. It is these
normal fields which allow us to push the $(n-1)$-manifolds off themselves
and again an extra factor of 2 in the value of linking pairings (e.g.
$\ell(x,x)$ is well defined in $\mathbb{Q}/2\mathbb{Z}$ not $\mathbb{Q}/\mathbb{Z}$ using the normal field). This
extra factor of 2 is recorded in the quadratic refinement of linking q_f.

Again the fact that the new information, q_f, is quadratic in nature
and thus delicate only on the two torsion is important. We will again
present the argument only for s-nice normal maps. Deleting all references to the based structure transforms this argument into one valid
for nice, normal maps.

Let $f: M^{2n-1} \to N^{2n-1}$ be an s-nice, normal map. By section I.2,
there is a pairing

$$\ell: \text{Tor } K_{n-1}(f) \times \text{Tor } K_{n-1}(f) \longrightarrow \mathbb{Q}/\mathbb{Z} \otimes \Lambda.$$

It has the following properties.

II.2.0.
a) It is Λ-linear in the second variable.

b) $\ell(x,y) = (-1)^n \ell(y,x)^-$.

c) $\text{Ad}(\ell): \text{Tor } K_{n-1}(f) \to \text{Hom}_\Lambda (\text{Tor } K_{n-1}(f), \mathbb{Q}/\mathbb{Z} \otimes \Lambda)$ is an
isomorphism.

Furthermore, there is an exact sequence

$$0 \longrightarrow A_{n-1} \overset{\varphi}{\longrightarrow} F_{n-1} \longrightarrow \text{Tor } K_{n-1}(f) \longrightarrow 0$$

where A_{n-1} and F_{n-1} are based Λ-modules and in these bases φ is the diagonal matrix

$$\begin{pmatrix} n_1 & & & & & 0 \\ & \cdot & & & & \\ & & \cdot & & & \\ & & & \cdot & & \\ & & & & \cdot & \\ 0 & & & & & n_r \end{pmatrix} \quad , \quad n_i \in \mathbb{Z} - \{0\}.$$

To see that $\ell(x,y) = (-1)^n \ell(y,x)^-$ let V_x^{n-1} and V_y^{n-1} be based, simply connected, disjoint cycles in M representing x and y. Let $n_1 V_x = \partial C_x$ and $n_2 V_y = \partial C_y$ for based simply connected chains C_x and C_y.

$$0 = \partial (C_x \cdot C_y) = n_1 V_x \cdot C_y + (-1)^{n-1} n_2 \, C_x \cdot V_y$$

$$= n_1 \, V_x \cdot C_y + (-1)^{n-1} n_2 (V_y \cdot C_x)^-.$$

Thus

$$0 = \frac{1}{n_2} V_x \cdot C_y + (-1)^{n-1} \frac{1}{n_1} (V_y \cdot C_x)^-$$

$$\equiv \ell(x,y) + (-1)^{n-1} \ell(y,x)^- \bmod \mathbb{Z}.$$

The surgery obstruction is determined by Tor $K_{n-1}(f)$, ℓ, and the "quadratic refinement" of ℓ, $q_f \colon K_{n-1}(f) \to \mathbb{Q}/\mathbb{Z} \otimes Q_n$. The map q_f satisfies

II.2.1.
$$\begin{cases} \text{a)} & q_f(x + y) = q_f(x) + q_f(y) + \ell(x,y) \quad \text{in } \mathbb{Q}/\mathbb{Z} \otimes Q_n. \\ \text{b)} & q_f(x) + (-1)^n q_f(x)^- = \ell(x,x) \text{ in } \mathbb{Q}/\mathbb{Z} \otimes \Lambda. \\ \text{c)} & q_f(x \cdot a) = \bar{a} q_f(x) a \quad \text{for } a \in \Lambda. \\ \text{d)} & q_f(x) \text{ has a representative } \alpha \in \mathbb{Q} \otimes \Lambda \text{ such that } \alpha = (-1)^n \bar{\alpha}. \end{cases}$$

There is a map $\varphi \colon Q_n \to \Lambda$ defined by $\varphi([\alpha]) = \alpha + (-1)^n \bar{\alpha}$. It induces a map $1 \otimes \varphi \colon \mathbb{Q}/\mathbb{Z} \otimes Q_n \to \mathbb{Q}/\mathbb{Z} \otimes \Lambda$. II.2.1 b) means that $1 \otimes \varphi (q_f(x)) = \ell(x,x)$.

If we partition the elements of $\pi = \pi_1(N)$ into $\{g_i, g_i^{-1}\}$, $\{t_i\}$, and $\{s_i\}$ where $g_i \neq g_i^{-1}$, $t_i = t_i^{-1}$ and $w(t_i) = (-1)^n$, and $s_i = s_i^{-1}$ and

$w(s_i) = (-1)^{n+1}$, then

$$\ell(x,x) = \sum_{g_i} (\ell_{g_i} \cdot g_i + (-1)^n w(g_i) \ell_{g_i} \cdot g_i^{-1}) + \sum_{t_i} \ell_{t_i} \cdot t_i + \sum_{s_i} \ell_{s_i} \cdot s_i$$

where ℓ_{g_i} and ℓ_{t_i} are in $\mathbb{Q}/\mathbb{Z}$ and ℓ_{s_i} is in $\mathbb{Z}/2 \hookrightarrow \mathbb{Q}/\mathbb{Z}$. The existence of $q(x)$ such that $\ell(x,x) = q(x) + (-1)^n q(x)^-$ implies that all the ℓ_{s_i} are 0.

Using properties II.2.1 a)–d) for q_f we see that

$$q_f(x) = \sum_{g_i} (q_{g_i} \cdot g_i + (-1)^n w(g_i) \cdot q_{g_i} \cdot g_i^{-1}) + \sum_{t_i} q_{t_i} \cdot t_i$$

where $2q_{g_i} = \ell_{g_i}$, and $2q_{t_i} = \ell_{t_i}$ in $\mathbb{Q}/\mathbb{Z}$. In $Q_n \otimes \mathbb{Q}/\mathbb{Z}$,
$q_{g_i} \cdot g_i + (-1)^n w(g_i) q_{g_i} \cdot g_i^{-1} = 2q_{g_i} \cdot g_i = \ell_{g_i} \cdot g_i$. Thus

$$q_f(x) = [\sum_{g_i} \ell_{g_i} \cdot g_i + \sum_{t_i} q_{t_i} \cdot t_i] \quad \text{where} \quad 2q_{t_i} = \ell_{t_i},$$

and the new information in $q(x)$ is the division of ℓ_{t_i} by 2 in $\mathbb{Q}/\mathbb{Z}$. This is always possible in exactly two ways. Tor $K_{n-1}(f) \cong T_2 \oplus T_{odd}$ where T_2 is the nice submodule of elements of order a power of 2, and T_{odd} is the nice submodule of elements of odd order.

<u>Lemma</u> II.2.2: Any quadratic refinement, q_2, on T_2 satisfying II.2.1 a)–d) above has a unique extension to q on all of Tor $K_{n-1}(f)$ still satisfying II.2.1.

<u>Proof</u>: If $rx = 0$ for $r \in \mathbb{Z}$, then $0 = q(rx) = r^2 q(x)$. Hence, if $x \in T_{odd}$, the $q(x)$ is of odd order in $\mathbb{Q}/\mathbb{Z} \otimes \Lambda$. In $\mathbb{Q}/\mathbb{Z}$ there is a unique way to divide an element of odd order by 2 so that the result is also of odd order. Thus $q_{t_i}(x)$ must be this unique "$\frac{1}{2} \ell_{t_i}(x,x)$" which is of odd order. This proves that for $x \in T_{odd}$, $q(x)$ can have at most one value, "$\frac{1}{2} \ell(x,x)$". It is easy to show that this indeed gives a form q_{odd} satisfying II.2.1 a)–d). For an arbitrary $x \in$ Tor $K_{n-1}(f)$ we write

$x = x_2 + x_{odd}$ and define $q(x) = q_2(x_2) + q_{odd}(x_{odd})$. Since $\ell(x_2, x_{odd}) = 0$ this is the unique extension of q_2 satisfying II.2.1.

On T_2 q_f is not determined by II.2.1 and the linking pairing, ℓ. If $x \in T_2$, then there is a relative bordism element representing x,

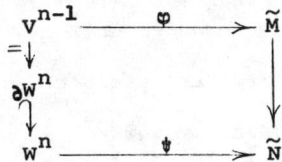

where φ is a based map and $\varphi_*[V] = x$. This follows since $K_{n-1}(f)$ $= H_n(\widetilde{N}, \widetilde{M})$, and $\Omega_n(\widetilde{N}, \widetilde{M}) \to H_n(\widetilde{N}, \widetilde{M})$ is onto two torsion. Given such a representative for x, we will define $q_f(x)$. Take any $(n-1)$-dimensional reduction of $\nu_W - \psi*(\xi)$. Restrict it to given an $(n-1)$ reduction of $\nu_V - \varphi*(\nu_{\widetilde{M}})$. This is equivalent to an embedding $V \underset{\varphi}{\hookrightarrow} \widetilde{M}$ and a nowhere zero normal field ϵ of $\nu(\varphi)$, [5]. We can assume that the projection of V into M is an embedding with normal field. Let V_ϵ be the "pushed off" copy of V along the field ϵ in M. Since $\varphi_*[V] = x$, some multiple of V bounds a chain C_V in $\widetilde{M}$, $\partial C_V = r \cdot V$. Let

$$q_f(x) = \frac{1}{2r}(C_V \cdot V_\epsilon) \quad \text{in} \quad \mathbb{Q}/\mathbb{Z} \otimes Q_n.$$

Theorem II.2.3: a) $q_f(x)$ defined is independent of all the choices made.

         q_f: Tor $K_{n-1}(f) \to (\mathbb{Q}/\mathbb{Z}) \otimes Q_n$ satisfies II.2.1 a)–d).

   b) Furthermore, we can do surgery on $f: M^{2n-1} \to N^{2n-1}$ to produce an s-nice normal map $f': M' \to N$ with $K_i(f') = 0$ for $i \neq n - 1$, and $K_{n-1}(f') = $ Tor $K_{n-1}(f)$ as modules with based structure, in such a way that the ℓ and q forms are unchanged.

   c) If f is a nice normal map, then so is f' and a) and b) are true, after omitting all references to the based structure.

Proof: b) and c) follow immediately from II.2.6 and the fact that all low dimensional surgeries can be done in the complement of the geometric data

which define ℓ and q_f. (The spheres on which we do surgery to kill

$K_{n-1}(f)$/Tor do not by general position miss the chains of dimension n

used to calculate ℓ and q_f. But, the argument in I.2.7 shows that we

can assume this by choosing our chains correctly.)

Suppose that we have done surgery until $f: M^{2n-1} \to N^{2n-1}$ satisfies

b). If $x \in K_{n-1}(f)$, then any two representatives for x, $(V^{n-1},W^n,\varphi,\psi)$

and $(V^{n-1'},W^{n'},\varphi',\psi')$ are bordant by some

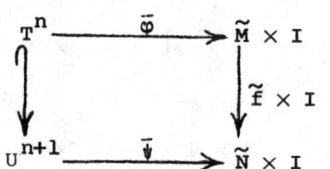

Given V^{n-1} embedded in M^{2n-1} with a normal field, ϵ, extend this to

an immersion of T^n in $M \times I$ with normal field. Restricting to the other

boundary component of T gives an embedding $V^{n-1'} \hookrightarrow M^{2n-1}$ with normal

field, ϵ'.

<u>Claim</u> II.2.4: If we use these two embedded manifolds $V \hookrightarrow M$ and $V' \hookrightarrow M$

with normal fields, ϵ and ϵ', as above to calculate $q_f(x)$, then they

give the same value.

<u>Proof:</u>

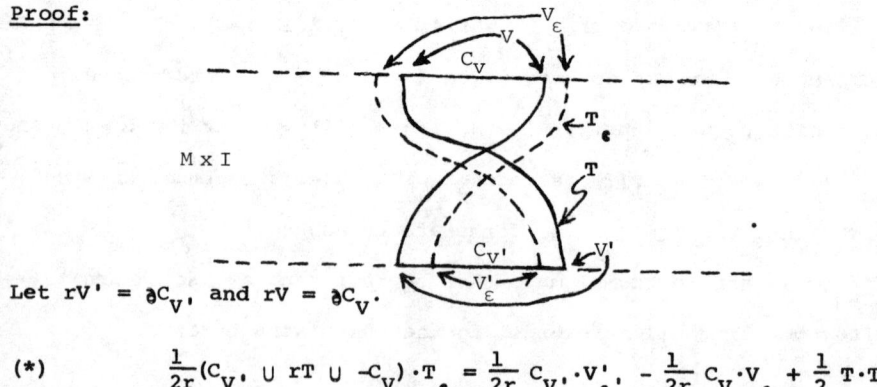

Let $rV' = \partial C_{V'}$ and $rV = \partial C_V$.

$$(*) \qquad \frac{1}{2r}(C_{V'} \cup rT \cup -C_V) \cdot T_\epsilon = \frac{1}{2r} C_{V'} \cdot V'_{\epsilon'} - \frac{1}{2r} C_V \cdot V_\epsilon + \frac{1}{2} T \cdot T_\epsilon .$$

Here $T \cdot T_\epsilon$ is the geometric self-intersection of T and T_ϵ. Since rT_ϵ is the image of a closed cycle in $\tilde{M} \times I$ and since the intersection of closed cycles in $\tilde{M} \times I$ is zero, the left hand side of (*) is 0 in $\mathbb{Q} \otimes \Lambda$. Thus we have

$$0 = \frac{1}{2r} C_{V'} \cdot V'_\epsilon - \frac{1}{2r} C_V \cdot V_\epsilon + \frac{1}{2} T \cdot T_\epsilon \quad \text{in } \mathbb{Q} \otimes \Lambda.$$

$T \cdot T_\epsilon = s(T) + (-1)^n \overline{s(T)}$ where $s(T)$ is the geometric self-intersection of T. $s(T)$ is an element in Q_n and is defined exactly as self-intersection of closed, immersed manifolds. For this, it is important that the boundary of T be embedded. In Q_n, $s(T) = (-1)^n \overline{s(T)}$. Hence in Q_n, $T \cdot T_\epsilon = 2s(T)$. Thus $\frac{1}{2} T \cdot T_\epsilon$ is 0 in $\mathbb{Q}/\mathbb{Z} \otimes Q_n$. This proves that

$$0 = \frac{1}{2r} C_{V'} \cdot V'_\epsilon - \frac{1}{2r} C_V \cdot V_\epsilon \quad \text{in} \quad \mathbb{Q}/\mathbb{Z} \otimes Q_n$$

or

$$\frac{1}{2r} C_{V'} \cdot V'_\epsilon = \frac{1}{2r} C_V \cdot V_\epsilon \quad \text{in} \quad \mathbb{Q}/\mathbb{Z} \otimes Q_n.$$

Claim II.2.5: Using any normal field for V' induced from a bundle reduction of $\nu_{W'} - \psi'^* \xi$ and using the normal field ϵ' for V' as above give the same value for $q_f(x)$.

Proof: Let $W'' = -T \cup -W$ and $\psi'': W'' \to \tilde{M}$ be $\overline{\varphi} \cup \psi$. Then, ϵ' is induced from a reduction of $\nu_{W''} - \psi''^* \xi$ over W''. We compare the difference of these two normal fields. For n even, the fields are homotopic. For n odd, their difference is an integer which, reduced modulo 2, is

$$\langle w_n (\nu_{W' \cup W''} - (\psi' \cup \psi'')^* \xi), [W' \cup W''] \rangle.$$

As in II.1.4 this is equal to

$$\langle v_n (\psi' \cup \psi'')^* \xi, [W' \cup W''] \rangle,$$

which, in turn, is 0 since ξ is fiber homotopy equivalent to $\nu_{N^{2n-1}}$. Thus, our two fields used in defining $q_f(x)$ differ by an even integer. But the value of $q_f(x)$ in $\mathbb{Q}/\mathbb{Z} \otimes Q_n$ depends only on the homotopy class of the normal field modulo 2. This proves $q_f(x)$ is well defined. It is easily seen to satisfy II.2.1 a) and c) on all of T. The proof of II.2.5 is completed using the following proposition.

<u>Proposition</u> II.2.6: If V^{n-1} is embedded in M^{2n-1} with normal field ϵ induced from bundle data covering a normal map (as in the definition of $q_f(x)$), and $n_1 V = \partial C$, then

$$C \cdot V_\epsilon = (-1)^n \ \overline{C \cdot V_\epsilon}.$$

<u>Proof</u>: We can assume that all n_1 sheets of C come into V from the direction $-\epsilon$, and that, except at its boundary C is transverse to V.

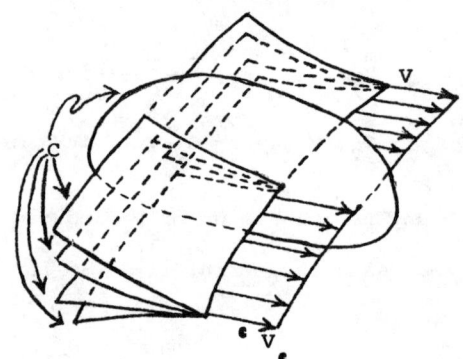

Let $C_\epsilon = C \cup n_1 E$ where E is $V \times I$ embedded along $+\epsilon$. $\partial(C_\epsilon) = n_1 V_\epsilon$. We shift C_ϵ transverse to V (keeping ∂C_ϵ fixed) by moving the n_1 copies of V in the (n-1) dimensional bundle perpendicular to ϵ in $\nu_{V \subset M}$, ζ^{n-1}. We see that $C_\epsilon \cdot V = C \cdot V_\epsilon + n_1 \chi(\zeta)$. Since ζ^{n-1} extends to a bundle over some W^n with $V = \partial W^n$, $\chi(\zeta^{n-1}) = 0$, and thus $C_\epsilon \cdot V = C \cdot V_\epsilon$. Now shift C_ϵ transverse to C without moving ∂C_ϵ or ∂C. We have

$$0 = \partial(C \cdot C_\epsilon) = rC \cdot V_\epsilon - (-1)^n r V \cdot V_\epsilon$$

$$= r(C \cdot V_\epsilon - (-1)^n \ \overline{C_\epsilon \cdot V})$$

$$= r(C \cdot V_\epsilon - (-1)^n \ \overline{C \cdot V_\epsilon}).$$

Thus,

$$C \cdot V_\epsilon = (-1)^n \overline{C \cdot V_\epsilon}.$$

Property II.2.1 d) follows immediately since $q_f(x) = [\frac{1}{2n_1} C \cdot V_\epsilon]$ in $(\mathbb{Q}/\mathbb{Z}) \otimes Q_n$. Also

$$q_f(x) + (-1)^n \overline{q_f(x)} = \frac{1}{2n_1}(C_V \cdot V_\epsilon + (-1)^n \overline{C_V \cdot V_\epsilon}) = \frac{1}{2n_1}(2C_V \cdot V_\epsilon)$$

$$= \frac{1}{2n_1} C_V \cdot V_\epsilon = \ell(x,x).$$

This gives II.2.1 b) and completes the proof of II.2.5.

Though the relation between the triples (Tor $K_{n-1}(f), \ell, q_f$) and Wall's surgery obstruction groups is not required in the sequel, we include a sketch of the following.

Theorem II.2.7: a) (Tor $K_{n-1}(f), \ell, q_f$) determines the Wall surgery
            obstruction in $L_{2n-1}(\pi)$ if f is nice;

        b) If f is s-nice, then (Tor $K_{n-1}(f), \ell, q_f$) together with the free,
            based resolution of Tor $K_{n-1}(f)$, determines the Wall surgery
            obstruction in $L_{2n-1}^s(\pi)$.

(See [15] page 56 for the definition of the odd Wall groups.)

Sketch of Proof: Let $a_1, \ldots, a_\ell$ be a natural generating set for Tor $K_{n-1}(f)$ with $n_i =$ order a_i.

Step I: $\exists \{\alpha_{ij}\}$, $\alpha_{ij} \in \Lambda$ such that
    1) $n_j \alpha_{ij} = (-1)^n n_i \overline{\alpha_{ji}}$
    2) $\frac{1}{n_i} \alpha_{ij} = \ell(a_i, a_j)$
    d) $\frac{1}{2n_i} \alpha_{ii} = q_f(a_i)$ in $\mathbb{Q}/\mathbb{Z} \otimes Q_n$.

Step II: Let $H_{n-1} = \Lambda(e_1, \ldots, e_\ell, f_1, \ldots, f_\ell)$ equipped with the intersection and self-intersections of the $(-1)^{n-1}$ symmetric hyperbolic form. For any collection of $\{\alpha_{ij}\}$ as in step I, define $K \subset H_{n-1}$ to be the based subspace with basis

$$\{n_1 e_1 + \sum_{j=1}^{\ell} \alpha_{1j} f_j, \ldots, n_\ell e_\ell + \sum_{j=1}^{\ell} \alpha_{\ell j} f_j\}.$$

$K \hookrightarrow H_{n-1}$ is a based subkernel. For any choice of $\{\alpha_{ij}\}$ as in step I, the class K determines in $L_{2n-1}^S(\pi)$ is the surgery obstruction of f. The main point in the proof of this is to show any set of $\{\alpha_{ij}\}$ as in step I is realized as the chain intersection matrix for a set of embedded spheres with trivial normal bundles $\{S_1^{n-1}, \ldots, S_\ell^{n-1}\}$ and chains C_i^n with $\partial C_i = n_i S_i^{n-1}$. See proposition II.3.3 for a proof of this.

Section II.3 - Forms representing the trivial obstruction. In sections II.1 and II.2 we gave geometrically defined algebraic pairings associated to a surgery problem $f: M^n \to N^n$, and showed that these pairings algebraically determined the surgery obstruction $\sigma(f) \in L_n^S(\pi_1(N))$. In this section, we examine which algebraic pairings determine the 0 element in $L_n^S(\pi)$. We find necessary and sufficient algebraic conditions in general for a pairing to define the zero element. Since our algebraic pairings for n even agree with those Wall uses to define $L_n^S(\pi)$, we use his conditions for this case.

II.3.1: Even dimensional case, $n = 2k$: Let (G, λ, μ) be as in section 4. It determines 0 in $L_{2k}^S(\pi)$ if and only if there is a based submodule $K \subset G$ such that

 1) $\lambda | K \times K = 0$,

 2) $\mu | K = 0$, and

 3) $\mathrm{Ad}(\lambda): K \to \mathrm{Hom}_\Lambda(G/K, \Lambda)$ is a simple isomorphism.

If G is not based, then (G, λ, μ) determines zero in $L_{2n}(\pi)$ if and only if there is a free subspace $K \hookrightarrow G$ such that 1), 2), and 3) (with "simple" deleted) hold.

Such a subspace, K, is called a subkernel for (G,λ,μ). For a proof
that the existence of such a $K \subset G$ is necessary and sufficient for (G,λ,μ)
to be   0   in $L_{2k}(\pi)$ (or $L_{2k}^s(\pi)$) see [15], page 47.

In the odd dimensional case our formalism is different from Wall's.
We develop a necessary and sufficient algebraic condition on the linking
pairing and its quadratic refinement for doing surgery on a nice (or
s-nice) normal map to produce a homotopy equivalence (or simple homotopy
equivalence). We prove that the condition is sufficient by actually doing
the surgery, not by making an algebraic connection with Wall's formalism.

As with the analysis of surgery on torsion classes outside the
middle dimension, here also the analysis is made in terms of a based
Λ-resolution for Tor $K_{n-1}(f)$

$$0 \longrightarrow A_{n-1} \longrightarrow F_{n-1} \longrightarrow \text{Tor } K_{n-1}(f) \longrightarrow 0.$$

We study the chain intersection pairing which induces a map

$$I \colon F_{n-1} \times F_{n-1} \longrightarrow \mathbb{Q} \otimes \Lambda$$

which resolves the linking pairing and its quadratic refinement on
Tor $K_{n-1}(f)$. The key property to be able to do surgery is that
$I | A_{n-1} \times F_{n-1} \to \mathbb{Z} \otimes \Lambda$ be non-singular. In the case of linking between
different modules we found that it was always possible to pick resolutions
so that the chain intersection map is non-singular. In the case of self-
linking, this is not always possible, and the inability to do it is the
obstruction to performing surgery on a odd dimensional, nice, normal map
to produce a homotopy equivalence.

The first step is to find out what properties the chain intersection
map has, and then to show that any algebraic map with these properties can
be realized as the intersection pairing of an appropriate set of chains
and cycles. Finally, we show that surgery is possible when
$I \colon A_{n-1} \times F_{n-1} \to \Lambda$ is non-singular.

Let $f: M^{2n-1} \to N^{2n-1}$ be an s-nice normal map with

$\ell: \text{Tor } K_{n-1}(f) \times \text{Tor } K_{n-1}(f) \to \mathbb{Q}/\mathbb{Z} \otimes \Lambda$ the intersection pairing and

$q_f: \text{Tor } K_{n-1}(f) \to \mathbb{Q}/\mathbb{Z} \otimes Q_n$ the quadratic refinement.   Let

$0 \to A_{n-1} \overset{\rho}{\to} F_{n-1} \overset{\phi}{\to} \text{Tor } K_{n-1}(f) \to 0$ be the based Λ-resolution of $\text{Tor } K_{n-1}(f)$.

If f is a nice, normal map pick any resolution for $\text{Tor } K_{n-1}(f)$.   In

either case the matrix for ρ is

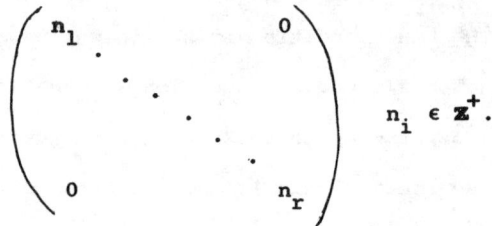

$n_i \in \mathbb{Z}^+.$

Suppose in addition, that our normal map is (n-1) connected. Then, we can

find a chain realization of the resolution such that each element in F_{n-1}

is represented by an embedded submanifold equipped with normal field

$z \in F_{n-1} \mapsto (Z^{n-1}, \epsilon_Z)$, and each element of A_{n-1} is represented by an

n-chain with correct boundary; $a \in A_{n-1} \mapsto C_a^n$.   We denote by Z' the copy of

Z pushed off along the normal field ϵ_Z.   Let $a = j^{-1}(Nz_1)$ and define

$$I(z_1, z_2) = \frac{1}{N}(C_a \cdot Z_2').$$

This, of course, is the unique extension of the chain intersection map

$$A_{n-1} \times F_{n-1} \longrightarrow \Lambda \quad \text{to a} \quad \Lambda - \text{map} \quad F_{n-1} \times F_{n-1} \longrightarrow \mathbb{Q} \otimes \Lambda.$$

<u>Proposition</u> II.3.2:

1)   I is Λ-linear in the second variable and Λ-anti-linear in the

   first,

2) $I(x,y) = (-1)^n I(y,x)^-,$

3) $I|: A_{n-1} \times F_{n-1} \to \mathbb{Z} \otimes \Lambda \hookrightarrow \mathbb{Q} \otimes \Lambda,$

4)   I induces the linking pairing on $\text{Tor } K_{n-1}(f)$ in that $I(x,y)$

   is a representative for $\ell(\phi(x), \phi(y))$, and

5)   I induces q_f on Tor $K_{n-1}(f)$ in that $\frac{1}{2} I(x,x)$ in $\mathbb{Q} \otimes \Lambda$ is a

representative for $q_f(\varphi(x))$.

<u>Proof</u>: 1), 3), 4), and 5) follow immediately from the definitions, 2) is

a consequence of II.2.6.

We also need a converse to this which tells us that all such algebraic

pairings are realized by the intersections of appropriate chains and

cycles. This converse is the analogue of lemma I.2.2 in the middle

dimension.

<u>Proposition</u> II.3.3: Let $f: M^{2n-1} \to N^{2n-1}$ be an $(n-1)$ connected, s-nice,

normal map with $K_{n-1}(f) = $ Tor $K_{n-1}(f)$ and with pairings ℓ and q_f and

resolution $0 \to A_{n-1} \to F_{n-1} \overset{\varphi}{\to} $ Tor $K_{n-1}(f) \to 0$ as before. We call

$\{z_1, \ldots, z_r\}$ the basis for F_{n-1} and suppose $\{n_1 z_1, \ldots, n_r z_r\}$ is the basis

for A_{n-1}. For any pairing $I: F_{n-1} \times F_{n-1} \to \mathbb{Q} \otimes \Lambda$ satisfying II.3.1, 1) –

5) there are disjointly embedded spheres with trivialized normal bundles

coming from the bundle map covering f

$$\{S_i^{n-1} \hookrightarrow M^{2n-1}\}$$

and manifolds $C_i^n \to M^{2n-1}$ such that

a) $[S_i^{n-1}] = \varphi(z_i)$,

b) $n_i S_i = \partial C_i$ in M, and

c) if we denote S_i' the "push off" of S_i along the first normal field

of the trivialization of its normal bundle, then

$$\frac{1}{n_i} C_i \cdot S_j' = I(z_i, z_j) \quad \text{in} \quad \mathbb{Q} \otimes \Lambda .$$

<u>Proof</u>: Find disjointly embedded spheres $\{S_1^{n-1}, \ldots, S_r^{n-1}\}$ with trivial

normal bundle representing $\{z_j, \ldots, z_r\}$ with $n_i S_i = \partial C_i^n$. We have two

pairings

$$I, I': F_{n-1} \times F_{n-1} \longrightarrow \mathbb{Q} \otimes \Lambda$$

satisfying II.3.2. I is the one we want, and I' is the one produced by
the spheres we have. We will modify the spheres until I = I'. By conditions 1 and 2, it suffices to have

$$I'(z_i,z_j) = I(z_i,z_j) \quad \text{for} \quad i \le j.$$

By 4, $I(z_i,z_j) - I'(z_i,z_j) \in \mathbb{Z} \otimes \Lambda$. Moving S_j through an isotopy which
intersects S_i transversally in $\pm g \in \Lambda$ and misses all other S_k, and adjoin-
ing n_j copies of the track of this isotopy to C_j changes $I'(z_i,z_j)$ by
$\pm g$ and leaves all other $I'(z_k,z_\ell)$ unchanged (except $I'(z_j,z_i)$). By a
sequence of these changes we can make $I(z_i,z_j) = I'(z_i,z_j)$ for $i < j$.
Changing the embedding $S_j^{n-1} \hookrightarrow M^{2n-1}$ with normal field ϵ_j by a regular
homotopy with normal field extending ϵ_j and with self intersection α in
$M^{2n-1} \times I$ changes $\frac{1}{n_j} C_j \cdot S_j$ by $\alpha + (-1)^n \bar{\alpha}$.

By condition 5) of II.3.1 we know that $\frac{1}{2} I(z_i,z_i) = \frac{1}{2} I'(z_i,z_i)$ in
$\mathbb{Q}/\mathbb{Z} \otimes \mathbb{Q}_n$. A straightforward, algebraic calculation shows that if an
element, $a \in \mathbb{Q} \otimes \Lambda$ satisfies $\frac{1}{2} a = 0$ in $\mathbb{Q}/\mathbb{Z} \otimes \Lambda$ and $a = (-1)^n \bar{a}$, then
$a = \alpha + (-1)^n \bar{\alpha}$ for some $\alpha \in \Lambda$. Applying this to $I(z_i,z_i) - I'(z_i,z_i)$,
one shows that by varying the embedding of S_i^{n-1} by a regular homotopy we
can make $I(z_i,z_i) = I'(z_i,z_i)$.

The change in the chain realization required to make I' = I was done
entirely by homotopies of chains. Thus, if the original chain realization
induced a simply homotopy equivalence with $C_*(f)$, then the new chain
realization does also.

<u>Theorem</u> II.3.4: Let f: $M^{2n-1} \to N^{2n-1}$ be an s-nice (nice) normal map.
If Tor $K_{n-1}(f)$ admits a resolution $0 \to A_{n-1} \to F_{n-1} \to$ Tor $K_{n-1}(f) \to 0$ with
a pairing I: $F_{n-1} \times F_{n-1} \to \mathbb{Q} \otimes \Lambda$ satisfying II.3.2 and with
I|: $A_{n-1} \times F_{n-1} \to \Lambda$ non-singular, then f is normally bordant to a simple
homotopy equivalence (homotopy equivalence).

<u>Proof</u>: We can assume that f is $(n-1)$ connected and that $K_{n-1}(f)$ = Tor $K_{n-1}(f)$. Embed disjoint spheres $S_i^{n-1} \hookrightarrow M^{2n-1}$ which represent the basis $\{z_1, \ldots, z_t\}$ for F_{n-1}. There are chains C_i^n in M with $\partial C_i^n = n_i S_i^{n-1}$. The normal bundle of each S_i^{n-1} is trivialized by the bundle data covering f. Denote by S_i' a copy of S_i pushed off along the first vector in the trivialization of the normal bundle. By II.3.3 we can choose the embedded spheres so that $C_i \cdot S_j' = I(n_i z_i, z_j)$.

Let $G: W^{2n} \to N^{2n-1} \times I$ be the trace of surgery along the $\{S_i^{n-1}\}$, and $f': M' \to N$ be the "other end" of W. From the exact sequence of the pair (G, f) we find that

$$K_*(G) = \begin{cases} 0 & * \neq n \\ \\ A_{n-1} & * = n. \end{cases}$$

(This is an isomorphism of based Λ-modules.) The cycles in W representing the basis for $K_n(G)$ are $(-C_1 \cup n_1 d_1, \ldots, -C_t \cup n_t d_t$ where d_j is the handle added along S_j^{n-1}.

$$K_*(G, f') = \begin{cases} 0 & * \neq n \\ \\ F_{n-1}^* & * = n. \end{cases}$$

(Again, this is an isomorphism of based modules.) The long exact sequence of the pair (G, f') is

$$0 \longrightarrow K_n(f') \longrightarrow K_n(G) \xrightarrow{i_*} K_n(G, f') \longrightarrow K_{n-1}(f') \longrightarrow 0$$

with vertical isomorphisms

$$A_{n-1} \xrightarrow{i_*} F_{n-1}^*$$

<u>Claim</u>: $i_*: K_n(G) \to K_n(G, f')$ is $-ad(I)$.

<u>Proof of claim</u>: The map $K_n(G) \to K_n(G, f')$ is the adjoint of the intersection map $K_n(G) \times K_n(G, f) \to \Lambda$. The element $a_i \in A_{n-1}$ is represented by $-C_i \cup n_i \cdot d_i$, and $z_j \in F_{n-1}$ is represented by $d_j \cup S_j^{n-1} \times I$. The

intersection $a_i \cdot z_j = -C_i \cdot S'_j = -I(a_i, z_j)$.

Since $\text{ad}(I|)$ is a simple isomorphism, f' is a simple homotopy equivalence.

Based torsion Λ-modules come equipped with a simple equivalence class of short free resolutions and thus we can view a based torsion module as the homology of a based chain complex. Consequently it makes sense to say that a short exact sequence of based torsion Λ-modules has trivial Whitehead torsion.

Theorem II.3.5: Suppose $f: M^{2n-1} \to N^{2n-1}$ is an s-nice (nice) normal map. If there is a nice submodule $T \hookrightarrow \text{Tor } K_{n-1}(f)$ on which ℓ and q_f vanish identically and so that

$$T \xrightarrow{ i } \text{Tor } K_{n-1}(f) \xrightarrow{ i* } T*$$

is a short exact sequence with trivial Whitehead torsion, then it is possible to do surgery on f to make it a simple homotopy equivalence (homotopy equivalence). We call such a submodule $T \hookrightarrow \text{Tor } K_{n-1}(f)$ a subkernel.

Proof: We can assume that $K_i(f) = 0$ for $i \neq n-1$ and that $K_{n-1}(f) = \text{Tor}(K_{n-1}(f))$. Let $A \to F \to T$ be the based free resolution for T. Find inside M^{2n-1} disjointly imbedded spheres $\{S_i^{n-1}\}$ representing the basis for F. Representing the basis for A there are chains C_i^n with $\partial C_i^n = n_i S_i^{n-1}$. As before we can arrange that $C_i^n \cdot S'_j = 0 \ \forall i, j$. This is because $\ell | T \times T = 0$ and $q_f | T = 0$. Now do surgery on these spheres. Let $G: W^{2n} \to N \times I$ be the trace and $f': M' \to N$ be the "other end".

Claim:
$$K_*(G) = \begin{cases} 0 & * \neq n-1, n \\ T* & * = n-1 \\ A & * = n \end{cases} \qquad \text{as based } \Lambda \text{ modules.}$$

Proof: The exact sequence of the pair (G, f) is

$$K_n(f) \longrightarrow K_n(G) \longrightarrow K_n(G,f) \longrightarrow K_{n-1}(f) \longrightarrow K_{n-1}(G) \longrightarrow 0$$

with vertical maps: $=$ down to 0 under $K_n(f)$; $=$ down under $K_n(G,f)$ to $F \longrightarrow \text{Tor } K_{n-1}(f)$; $=$ down under $K_{n-1}(f)$.

__Claim:__

$$K_*(f') = \begin{cases} 0 & * \neq n-1, n \\ A^* & * = n-1 \\ A & * = n \end{cases} \qquad \text{as based } \Lambda\text{-modules .}$$

__Proof:__ The exact sequence for (G, f') is

$$0 \longrightarrow K_n(f') \longrightarrow K_n(G) \longrightarrow K_n(G,f') \longrightarrow K_{n-1}(f') \longrightarrow K_{n-1}(G) \longrightarrow 0$$

with vertical maps: $=$ down under $K_n(G)$ to $A \xrightarrow{\text{ad(inter.)}} F^*$; $=$ down under $K_n(G,f')$; and $=$ down under $K_{n-1}(G)$ to T^*.

But the intersection $A \times F \to \Lambda$ is identically 0. Thus $K_n(f') = A$ and $K_{n-1}(f')$ sits in an exact sequence

$$0 \longrightarrow F^* \longrightarrow K_{n-1}(f') \longrightarrow T^* \longrightarrow 0.$$

In M^{2n-1} the generating set for T^* is represented by disjointly embedded spheres of dimension $(n-1)$, $\{S_i'\}$. We can choose the spheres so that $n_i S_i'$ bounds a chain c_i^n in M^{2n-1} with $c_i^n \cdot c_j^{n-1} = \delta_{ij}$. The spheres S_i' sit naturally in M' and c_i^n-(neighborhood $c_i \cdot S_i$) is a homology in M' from $n_i S_i'$ to the basis for F^*. This proves that $K_{n-1}(f') = A^*$. The intersection pairing $K_{n-1}(f') \times K_n(f') \to \Lambda$ is the natural non-singular one

$$A^* \times A \longrightarrow \Lambda.$$

Surgery on the basis for $A^* = K_{n-1}(f')$ produces a normal bordism from f' to a simple homotopy equivalence, (see I.3.4).

CHAPTER III: The Index and the de Rham Invariant

In this chapter we study the homology of closed, oriented simply connected manifolds. In the spirit of the previous sections, we concentrate on that part of the homology which is paired with itself by Poincare duality, i.e. $H_k(L^{2k})/\text{Tor}$ and $\text{Tor } H_k(L^{2k+1})$. We analyse these self-pairings algebraically and find two invariants of interest--the index in the case of symmetric intersection and the de Rham invariant in the case of skew-symmetric linking. Both these invariants are classical, and the index, in particular, has been much studied. In the case of linking pairings we also discuss chain realizations for the pairings.

Theorem III.1: If k is odd, then the intersection pairing $H_k(L^{2k})/\text{Tor} \otimes H_k(L^{2k})/\text{Tor} \to \mathbb{Z}$ is skew-symmetric and non-singular over $\mathbb{Z}$. Algebraically it is isomorphic to a direct sum of $\{\mathbb{Z} \oplus \mathbb{Z}, \begin{pmatrix} 0 & 1 \\ -1 & 0 \end{pmatrix}\}$.

Proof: That the pairing is skew symmetric and non-singular is classical. The algebraic classification of such pairings is straightforward, see [9].

If k is even denote $H_k(L^{2k})/\text{Tor}$ by $F_k(L^{2k})$. The pairing

$$F_k(L) \otimes F_k(L) \longrightarrow \mathbb{Z}$$

is symmetric and non-singular. Let r be the rank of $F_k(L)$, and let d be the rank of a maximal subspace $K \hookrightarrow F_k(L)$ on which the pairing vanishes. The number d is independent of the maximal subspace chosen, and the pairing $(F, \cdot)$ decomposes as

$$(A, \cdot) \oplus (B, \cdot)$$

where $(A, \cdot) \cong \oplus(\mathbb{Z} \oplus \mathbb{Z}, \begin{pmatrix} 0 & 1 \\ 1 & * \end{pmatrix})$ and $(B, \cdot)$ is $\pm$ definite. The signature of

$(F, \cdot)$ is defined to be $\pm$ rk B depending on where B is $\pm$ definite. The signature is an additive invariant under orthogonal direct sum, see [7].

<u>Theorem</u> III.2: If k is even, then the signature of L^{2k} is zero if and only if there is a subspace $K \hookrightarrow F_k(L^{2k})$ satisfying

1) $\cdot | K \otimes K$ vanishes, and

2) $\text{ad}(\cdot): K \to (F_k(L)/K)^*$ is an isomorphism.

The signature of the pairing is called the index of L, $I(L)$.

<u>Theorem</u> III.3: If $2k + 1 \equiv 3(4)$, then the linking pairing

$$\ell: \text{Tor } H_k(L^{2k+1}) \otimes \text{Tor } H_k(L^{2k+1}) \longrightarrow \mathbb{Q}/\mathbb{Z}$$

is non-singular and symmetric. Thus it admits a "resolution". That is there is a free abelian group F, an epimorphism $F \overset{\pi}{\to} \text{Tor } H_k(L^{2k+1})$, and a symmetric pairing $F \otimes F \overset{I}{\to} \mathbb{Q}$ such that

1) $I | (\ker \pi) \otimes F$ takes values in $\mathbb{Z}$,

2) $I(x,y) = \ell(\pi(x), \pi(y))$ modulo $\mathbb{Z}$, and

3) $\text{Ad}(I)$ induces an isomorphism $\text{Ker } \pi \to F^* = \text{Hom}(F, \mathbb{Z})$.

<u>Proof</u>: Given any free abelian group mapping onto $\text{Tor } H_k(L^{2k+1})$ there is a pairing satisfying 1) and 2). The crucial property is 3). That such a resolution exists follows from [14], theorem 6.

<u>Theorem</u> III.4: If $2k + 1 \equiv 1(4)$, then the linking pairing $\text{Tor } H_k(L) \otimes \text{Tor } H_k(L) \to \mathbb{Q}/\mathbb{Z}$ is non-singular and skew-symmetric. This only means $\ell(x,x)$ is of order 2. In fact $\ell(x,x) = \langle v_k(v_L), x \rangle \in \mathbb{Z}/2 \subset \mathbb{Q}/\mathbb{Z}$. $\text{Tor } H_k(L)$ is isomorphic to $A \oplus A \oplus \epsilon \mathbb{Z}/2$ where $\epsilon = 0$ or 1. The pairing is a direct sum of pairings $\begin{pmatrix} 0 & \frac{1}{n} \\ -\frac{1}{n} & * \end{pmatrix}$ on $\mathbb{Z}/n \oplus \mathbb{Z}/n$, and $\left(\frac{1}{2}\right)$ on $\mathbb{Z}/2$.

<u>Proof</u>: All this follows from the standard algebraic classification of skew-symmetric linking pairings, see [8]. The reason ϵ modulo 2 is all that is needed is that

$$\{\mathbb{Z}/2, (\tfrac{1}{2})\} \oplus \{\mathbb{Z}/2, (\tfrac{1}{2})\} = \left\{ \mathbb{Z}/2 \oplus \mathbb{Z}/2, \begin{pmatrix} 0 & \tfrac{1}{2} \\ -\tfrac{1}{2} & \tfrac{1}{2} \end{pmatrix} \right\}.$$

The element $\epsilon \in \mathbb{Z}/2$ is the <u>de Rham Invariant</u> of the pairing and of the manifold L^{2k+1}. It is denoted $d(L)$.

On the chain level 4 has the following consequence. If $d(L) = 0$, then there are cycles $\{z_1^k, \ldots, z_{2r}^k\}$ and chains $\{c_1^{k+1}, \ldots, c_{2r}^{k+1}\}$ with

1) $\partial c_i = n_i z_i$,

2) $n_{i+r} = n_i$,

3) the cycles $\{z_i\}$ induce an isomorphism $\overset{2r}{\underset{i=1}{\oplus}} \mathbb{Z}/n_i \mathbb{Z} \overset{\cong}{\to} \text{Tor } H_k(L)$, and

4) there are pushed off copies of Z_i, Z_i' with the chain intersection matrix given by

$$\begin{pmatrix} 0 & \begin{pmatrix} \tfrac{1}{n_1} & & 0 \\ & \ddots & \\ 0 & & \tfrac{1}{n_r} \end{pmatrix} \\ \begin{pmatrix} -\tfrac{1}{n_1} & & 0 \\ & \ddots & \\ 0 & & -\tfrac{1}{n_r} \end{pmatrix} & * \end{pmatrix}.$$

Both the de Rham invariant and the index are bordism invariants. If $M^{4k} = \partial W^{4k+1}$, then the kernel of $(H_{2k}(M)/\text{Tor} \to H_{2k}(W)/\text{Tor})$ provides a subspace K with $(\cdot)_M | K \otimes K = 0$, and $\text{Ad}(\cdot_M): K \to [H_{2k}(M)/\text{Tor} + K]^*$ an isomorphism. This implies $I(M) = 0$. The de Rham invariant of L^{4k+1} is measured by the characteristic class

$$\langle v_{2k} \cdot \text{Sq}^1 v_{2k}, [L] \rangle \in \mathbb{Z}/2\mathbb{Z}$$

where v_{2k} is the $2k^{\text{th}}$ Wu class, see [9], $(\langle v_{2k}, [z_i^{2k}] \rangle = \ell([z_i], [z_i]) \in \mathbb{Z}/2\mathbb{Z})$.

There is also a direct Poincaré

duality proof that the de Rham invariant of a boundary is 0. The de Rham invariant is multiplicative with respect to the index:

$$d(L^\ell \times M^m) = \begin{cases} d(L) \cdot I(M) & \ell \equiv 1, \ m \equiv 0 \ (4) \\ I(L) \cdot d(M) & \ell \equiv 0, \ m \equiv 1 \ (4) \\ 0 & \text{otherwise} \end{cases}$$

There is a 5-manifold of de Rham invariant 1. Let $c: \mathbb{C}P^2 \to \mathbb{C}P^2$ be complex conjugation. Then $M^5 = \mathbb{C}P^2 \times I/\{(X,0) \sim (c(X),1)\}$ has $H_2(M^5) = \mathbb{Z}/2$. Thus $d(M^5) = 1$. M^5 is not simply connected ($\pi_1 = \mathbb{Z}$), but we can do a one dimensional surgery to replace M^5 by a simply connected manifold X^5. X^5 is diffeomorphic to $SU(3)/SO(3)$ and has de Rham invariant 1.

$\mathbb{C}P^2$ is the simplest example of a manifold of index 1.

CHAPTER IV: The Product Formula

<u>Section IV.1 - Even dimensional normal maps</u>. We are now in a position to
apply the analysis in chapter I through III to prove our produce formulae.
Given a normal map f: $M^n \to N^n$ with $f|\partial M$ a simple homotopy equivalence (or
homotopy equivalence), and a closed, oriented simply connected manifold
L^ℓ, we can form a new normal map

$$f \times 1_L: M^n \times L^\ell \longrightarrow N^n \times L^\ell.$$

The surgery obstruction of $f \times 1_L, \sigma(f \times 1_L) \in L_{n+\ell}^S(\pi_1(N))$ is easily seen to
depend only on the obstruction $\sigma(f) \in L_n^S(\pi_1(N))$ and the class of L in
oriented bordism, Ω_ℓ. Furthermore, $\sigma(f \times 1_L)$ is additive in both of these
factors. Thus, this process of crossing with a closed, simply connected
manifold induces homomorphisms

$$L_n(\pi) \otimes \Omega_\ell \xrightarrow{\omega} L_{n+\ell}(\pi)$$

and

$$L_n^S(\pi) \otimes \Omega_\ell \xrightarrow{\omega^S} L_{n+\ell}^S(\pi).$$

By a product formula, we mean an explicit formula for ω and ω^S.

<u>Theorem</u> IV.1.1: a) If $\ell \equiv 2(4)$ or $3(4)$, then $\omega: L_n(\pi) \otimes \Omega_\ell \to L_{n+\ell}(\pi)$
        and $\omega^S: L_n^S(\pi) \otimes \Omega_\ell \to L_{n+\ell}^S(\pi)$ are 0.

   b) if $\ell \equiv 0(4)$, then there is a natural periodicity identification
        of $L_n(\pi)$ with $L_{n+\ell}(\pi)$ and $L_n^S(\pi)$ with $L_{n+\ell}^S(\pi)$, see [15].
        With these identifications both ω and ω^S are multiplication by
        the index of L . Thus, for instance

$$
\begin{array}{ccc}
L_n^S(\pi) \otimes \Omega_{4\ell} & \xrightarrow{\quad \omega^s \quad} & L_n^S(\pi) \\
\downarrow{\scriptstyle 1 \otimes I} & & \parallel \\
L_n^S(\pi) \otimes \mathbb{Z} & \xrightarrow{\quad = \quad} & L_n^S(\pi)
\end{array}
$$

commutes

c) If $\ell \equiv 1(4)$ then there are maps

$$
\varphi: L_n(\pi) \longrightarrow L_{n+1}(\pi)
$$

and

$$
\varphi^s: L_n^S(\pi) \longrightarrow L_{n+1}^S(\pi).
$$

Both im φ and im φ^s consist of element of order ≤ 2.

$$
\omega([f: M \to N] \otimes L^{4\ell+1}) = \varphi([f: M \to N]) \cdot d(L)
$$

and

$$
\omega^s[(f: M \to N] \otimes L^{4\ell+1}) = \varphi^s([f: M \to N]) \cdot d(L).
$$

Thus we see that the only invariants of a simply connected, closed manifold which appear in the product formulae are the index and the de Rham invariant. Both of these are algebraic invariants associated to the dual pairings on the homology in the middle dimension.

In this section we prove this formula if we start with an even dimensional surgery problem and cross it with any closed, simply connected manifold. In the next two sections we deal with odd dimensional surgery problems cross any closed, simply connected manifold.

We begin with a normal map $(f,b_f): (M^{2n}, \nu_M) \to (N^{2n}, \xi)$. We do surgery on (f,b_f) until $K_i(f) = 0$ for $i \neq n$ and $K_n(f) = G$ a free Λ-module (with a simple equivalence class of bases if $f|\partial M$ is a simple homotopy equivalence). G has a non-singular intersection pairing $\lambda: G \times G \to \Lambda$ (In fact if $f|\partial M$ is a simple homotopy equivalence, then $\mathrm{ad}(\lambda): G \to \mathrm{Hom}_\Lambda(G,\Lambda)$ is a simple isomorphism.) and a self intersection form μ. The triple (G,Λ,μ) satisfies II.1.1 a)-f) and the element it

determines in $L_{2n}(\pi_1(N))$ (or $(L_{2n}^S(\pi_1)))$ is the Wall surgery obstruction
of (f, b_f). We denote this element $\sigma(f)$. For simplicity, we deal only
with the case when $f|\partial M$ is a simple homotopy equivalence, and we are
calculating the surgery obstruction of the product in $L_*^S(\pi)$. The obvious
deletions of references to based structures is all that is required in the
other case.

Cross with L^ℓ. By this we mean form the normal map

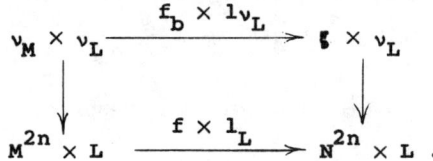

We denote this normal map by $f \times 1_L$. Restricted to $\partial(M \times L) = \partial M \times L$, it
is a simple homotopy equivalence. The first step in evaluating $\sigma(f \times 1_L)$
is to calculate the kernel modules and their pairings.

I: $K_i(f \times 1_L) = K_n(f) \underset{\mathbb{Z}}{\otimes} H_{i-n}(L)$ as Λ-module with based structure.

   (This follows from the Künneth theorem, since $K_n(f)$ is free.)

II: The intersection pairings $K_{n+i}(f \times 1_L)/\text{Tor} \times K_{n+\ell-i}(f \times 1_L)/\text{Tor}$
   $\to \Lambda$ are equal to $\lambda_f \otimes \cdot_L$. Here, $\cdot_L$, represents the usual intersection pairing in L.

III: The linking pairings, ℓ: Tor $K_{n+i}(f \times 1_L) \times$ Tor $K_{n+\ell-i-1}(f \times 1_L)$
   $\to (\mathbb{Q}/\mathbb{Z}) \otimes \Lambda$ are equal to $\lambda_f \otimes \ell_L$. Here ℓ_L is the usual linking
   pairing in L with values in $\mathbb{Q}/\mathbb{Z}$.

III follows by taking product cycle and chain representsions. For
$i \not\equiv \ell - i - 1$ let $0 \to A_i \to F_i \to$ Tor $H_i(L) \to 0$ and
$0 \to A_{n-i-1} \to F_{n-i-1} \to$ Tor $H_{n-i-1}(L) \to 0$ be integral resolutions. We can
assume that the chain intersection maps $A_i \otimes F_{n-i-1} \overset{\varphi_i}{\longrightarrow} \mathbb{Z}$ and
$A_{n-i-1} \otimes F_i \overset{\varphi_{n-i-1}}{\longrightarrow} \mathbb{Z}$ are non-singular. The tensor product of $K_n(f)$ with
these resolutions are resolutions for Tor $K_*(f \times 1_L)$. From III it
follows that the tensor product pairings $\lambda_f \otimes \varphi_i$ and $\lambda_f \otimes \varphi_{n-i-1}$ lift the

linking pairings. These are obviously simple isomorphisms. This together
with I and II proves that $f \times 1_L$ is an s-nice normal map.

We now begin the calculation of $\sigma(f \times 1_L)$. It is divided into the
case ℓ is even and the case ℓ is odd.

Subcase: ℓ is even, $\ell = 2k$.

IV.1.2: Since $f \times 1_L$ is s-nice, we need only calculate $\lambda_{f \times 1_L}$ and $\mu_{f \times 1_L}$
on $K_n(f) \otimes H_k(L)/\mathrm{Tor}$. We have seen that the intersection form is $\lambda_f \otimes \cdot_L$.

Claim: The μ-form for $f \times 1_L$ vanishes on $K_n(f) \otimes \mathrm{Tor}\ H_k(L)$ and on the
quotient $K_n(f) \otimes H_k(L)/\mathrm{Tor}$ it is determined by

$$\mu_{f \times 1_L}(x \otimes y) = \mu_f(x) \cdot (y \cdot y).$$

Proof: The proof consists of a local product formula. Given
$x \otimes y \in K_n(f) \otimes H_k(L)$, we represent x by an immersed sphere $S^n \rightsquigarrow M$ with
trivial normal bundle (the normal bundle reductions comes from a reduction
of $(\nu_{D^{n+1}} - \psi * \xi)$ for some $\psi: D^{n+1} \to N)$. We take this immersion to have
only transversal double points. We represent some odd multiple of y by
a manifold $Y^k \to L^{2k}$. By the Whitney trick, we embed $Y^k \hookrightarrow L^{2k}$. (If $k = 1$,
then $L = S^2$ and there is no H_1. If $k = 2$, the L is bordant to
$\pm(\mathbb{CP}^2 \# \ldots \# \mathbb{CP}^2)$ and a generating set of H_2 is represented by embedded two
spheres.) $S^n \times Y^k$ is then immersed in $M \times L$ representing $x \otimes y$ (or some
odd multiple). This is an "appropriate immersion" up to regular homotopy
for calculaing $\mu(x \otimes y)$ since the bundle reduction comes from one over
$D^{n+1} \times Y^\ell$. We must shift this immersion within its regular homotopy class
until it has only transverse double points. As it sits now, above each
double point of $S^n \rightsquigarrow M$, there is a double copy of Y^k in L^{2k}. Shift one of
these copies transverse to the other in the L^{2k} factor. We get
$\chi(\nu(Y \hookrightarrow L))$ points above each double point of S^n. Since $\chi(\nu(Y \hookrightarrow L)) = Y \cdot Y$,
we see that $\mu_{f \times 1_L}(x \otimes y) = \mu_f(x) \cdot (y \cdot y)$.

If $k \equiv 1(2)$, or $k \equiv 0(2)$ and $I(L^{2k}) = 0$, then there is in $H_k(L)/\text{Tor}$ a self-annihilating subspace K with $\text{Ad}(\cdot_L): K \to [(H_k(L)/\text{Tor})/K]^*$ an isomorphism. By the product formulae for $\lambda_{f \times 1_L}$ and $\mu_{f \times 1_L}$, we see that they both vanish on $K_n(f) \otimes K \hookrightarrow K_{n+k}(f \times 1_L)$. Also,

$$\text{Ad}(\lambda_{f \times 1_L}): K_n(f) \otimes K \longrightarrow [K_{n+k}(f \times 1_L)/K_n(f) \otimes K]^*$$

is identified with the isomorphism

$$\text{Ad}(\lambda_f) \otimes \text{Ad}(\cdot_L): K_n(f) \otimes K \longrightarrow K_n(f)^* \otimes [(H_k(L)/\text{Tor})/K]^*.$$

Applying II.3.1 shows that the surgery obstruction for $f \times 1_L$ is zero in $L^S_{2n+2k}(\pi)$.

If $I(L) \neq 0$, then $L^{2k} - a\underbrace{(\mathbb{C}P^2 \times \ldots \times \mathbb{C}P^2)}_{k/2 \text{ times}}$ has 0 index where $a = I(L)$. Thus the surgery obstruction $\sigma(f \times 1_{L-a(\mathbb{C}P^2 \times \ldots \times \mathbb{C}P^2)}) = 0$ or $\sigma(f \times 1_L) = a\sigma(f \times 1_{\mathbb{C}P^2 \times \ldots \times \mathbb{C}P^2})$. Thus to complete the proof of the product formula for

$$L^S_{2n}(\pi) \otimes \Omega_{2k} \xrightarrow{\omega^s} L^S_{2n+2k}(\pi),$$

it suffices to show that the surgery obstruction of $f \times 1_{\mathbb{C}P^2}$ equals that of f. This is just Wall's periodicity calculation. Since $H_2(\mathbb{C}P^2) = \mathbb{Z}$ and the intersection pairing is given by the matrix (1), the above calculations of the λ and μ forms for $f \times 1_{\mathbb{C}P^2}$ show that they are identical to the λ and μ forms for f. Hence, the obstructions are the same.

<u>Subcase</u> $\ell = 2k + 1$: Here we must calculate ℓ and $q_{f \times 1_L}$ on $\text{Tor } K_{n+k}(f \times 1_L) = K_n(f) \otimes \text{Tor } H_\ell(L)$. We have already seen that $\ell(x \otimes y, x' \otimes y') = \lambda_f(x,x') \otimes \ell(y,y')$ in $\Lambda \otimes \mathbb{Q}/\mathbb{Z}$. Since we know ℓ, it suffices to calculate $q_{f \times 1_L}$ on elements of the form $x \otimes y \in K_n(f) \otimes \text{Tor } H_\ell(L)$. This is our second local product formula.

If $x \in K_n(f)$, then $[\lambda_f(x,x)] \in \mathbb{Q}_n$ is divisible by 2. To see this let $\omega \in \Lambda$ be any element which projects to $\mu_f(x)$ in Q_n. Then

$\lambda_f(x,x) = \omega + (-1)^n \bar{\omega}$ in Λ.   In Q_n $[\lambda_f(x,x)] = [\omega] + [(-1)^n \bar{\omega}] = 2[\omega]$.

Let $[\frac{1}{2} \lambda_f(x,x)]$ denote any element in Q_n with the property that

$2[\frac{1}{2} \lambda_f(x,x)] = [\lambda_f(x,x)]$. If $\frac{r}{s}$ is a rational number then

$[\frac{1}{2} \lambda_f(x,x)] \otimes \frac{r}{s} \in Q_n \otimes \mathbb{Q}/\mathbb{Z}$ is independent of our choice of $[\frac{1}{2} \lambda_f(x,x)]$.

For if $2a = 2b = [\lambda(x,x)]$ in Q_n then

$$(a \otimes r/s) - (b \otimes r/s) = (a - b) \otimes r/s = 2(a - b) \otimes r/2s = 0.$$

Proposition IV.1.2: If $x \otimes y$ is an odd torsion element, then

$$
q_{f \times 1_L}(x \otimes y) = \begin{cases} [\frac{1}{2} \lambda_f(x,x)] \otimes \ell(y,y)] & \text{in } Q_{n+k+1} \otimes \mathbb{Q}/\mathbb{Z} \text{ if } \ell \equiv 3(4) \\ \\ 0 & \text{in } Q_{n+k+1} \otimes \mathbb{Q}/\mathbb{Z} \text{ if } \ell \equiv 1(4). \end{cases}
$$

Note: If $\ell \equiv 3(4)$, then $Q_{n+k+1} = Q_n$, and by the above discussion $[\frac{1}{2} \lambda_f(x,x)] \otimes \ell(y,y)$ is well defined.

Proof: This calculation is purely algebraic. Recall that $q_{f \times 1_L}(x \otimes y)$ is the unique element in $Q_{n+k+1} \otimes \mathbb{Q}/\mathbb{Z}$ which

    1)   is of odd order and

    2)   satisfies $q_{f \times 1_L}(x \otimes y) + (-1)^{n+k+1} q_{f \times 1_L}(x \otimes y)^- = \ell(x \otimes y, x \otimes y)$.

As we have seen $\ell(x \otimes y, x \otimes y) = \lambda_f(x,x) \otimes \ell(y,y)$. If $\ell \equiv 1(4)$, then $\ell(y,y)$ is of order 1 or 2. When y is odd torsion it must be 0. Thus $\ell(x \otimes y, x \otimes y) = 0$. Consequently $q_{f \times 1_L}(x \otimes y) = 0$ when $\ell \equiv 1(4)$ and y is odd torsion.

If $2k + 1 \equiv 3(4)$, then $\lambda_f(x,x) = (-1)^{n+k+1} \lambda_f(x,x)^-$. Thus

$[\frac{1}{2} \lambda_f(x,x)] \otimes \ell(y,y) + (-1)^{n+k+1} [\frac{1}{2} \lambda_f(x,x)]^- \otimes \ell(y,y) = \lambda_f(x,x) \otimes \ell(y,y) =$

$\ell_{f \times 1_L}(x \otimes y, x \otimes y)$. From the note following II.1.3 on the structure of

$\lambda(x,x)$, one checks that $[\frac{1}{2} \lambda_f(x,x)] \otimes \ell(y,y)$ is of odd order when $\ell(y,y)$ is.

Proposition IV.1.3: If $x \otimes y$ is of order a power of 2, then

$$q_{f \times 1_L}(x \otimes y) = \begin{cases} [\frac{1}{2} \lambda_f(x,x)] \otimes \ell(y,y)] \in Q_{n+k+1} \otimes \mathbb{Q}/\mathbb{Z} \text{ if } 2k+1 \equiv 3(4) \\[2em] \mu_f(x) \otimes \ell(y,y) \in Q_n \otimes \mathbb{Z}/2 = Q_{n+k+1} \otimes \mathbb{Z}/2 \hookrightarrow Q_{n+k+1} \otimes \mathbb{Q}/\mathbb{Z} \\[1em] \qquad\qquad\qquad\qquad\qquad\qquad\qquad \text{if } 2k+1 \equiv 1(4). \end{cases}$$

Note: $Q_n \otimes \mathbb{Z}/2 = Q_{n+k+1} \otimes \mathbb{Z}/2$ by definition. Since $\mathbb{Z}/2 \hookrightarrow \mathbb{Q}/\mathbb{Z}$ we have a map $Q_n \otimes \mathbb{Z}/2 \hookrightarrow Q_{n+k+1} \otimes \mathbb{Q}/\mathbb{Z}$.

Proof: In this case, the calculation is geometric and more delicate. We may assume, however, that x is represented by an immersed sphere $S^n \hookrightarrow M^{2n}$ with trivial normal bundle, and that y is represented by an embedded manifold $j: Y^k \hookrightarrow L^{2k+1}$ with $rY = \partial W^{k+1}$. Pick (arbitrarily) a nowhere zero normal field ϵ for Y in L. This gives a normal field for $S^n \times Y^k \hookrightarrow M^{2n} \times L^{2k+1}$. This normal field comes from the product bundle reduction $\epsilon^n \times \zeta$ of the normal bundle of $S^n \times Y$ in $M \times L$. (Here ζ is the perpendicular to the normal field ϵ of Y in L.) Consequently, <u>it is a correct normal field to calculate</u> $q_{f \times 1_L}(x \otimes y)$. Unfortunately, $S^m \times Y^k$ is not embedded in $M \times L$. Our first step is to change this immersion by a regular homotopy (carrying along the normal field) until it is embedded. Near each double point of $S^m \hookrightarrow M$ we have two sheets P_+ and P_- intersecting transversally in a point d.

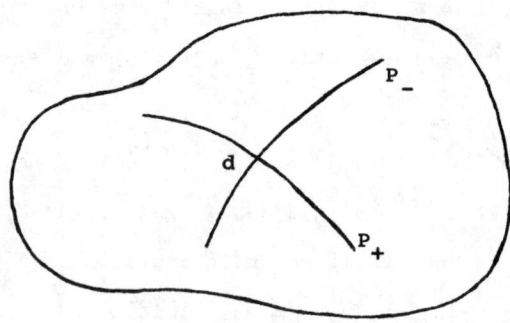

Above the double point there is a double copy of the manifold Y^k. Push the copy of Y_k above P_+ up the normal field ε, 2 units at d and damp

out the push above the rest of P_+

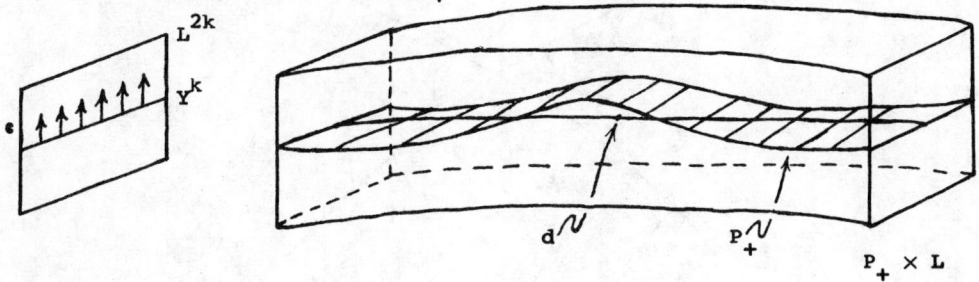

Doing this near each double point produces an embedding of

$S^n \times Y^k \underset{g}{\hookrightarrow} M \times L$ with normal field. We can extend the embedding g to a

map $\bar{g}\colon S^n \times W \to M \times L$.

We must intersect this with $g'\colon S^n \times Y^k \to M \times L$ where g' is the

result of a small push along the normal field ε from g. First push

$S^n \times W$ in the M direction along a normal field γ for $S^n \overset{\alpha}{\hookrightarrow} M$. (Recall

that $\nu(S^n \overset{\alpha}{\hookrightarrow} M)$ is trivial.) During this shift $p \times W$ moves through

$I \times W \hookrightarrow I \times L$ to $p' \times W$.

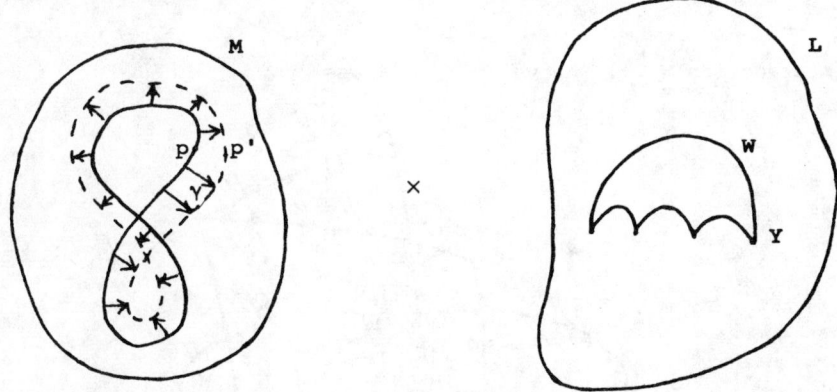

Consequently, the boundary of $S^n \times W$ does not intersect $g'(S^n \times Y)$ during

the shift, and thus the intersection number $(S^n \times W) \cdot g'(S^n \times Y)$ is

unchanged. Near each double point, d, there are two points p^1 and p^2

in M where the end points of γ intersect the original immersion $S^n \overset{\alpha}{\hookrightarrow} M$.

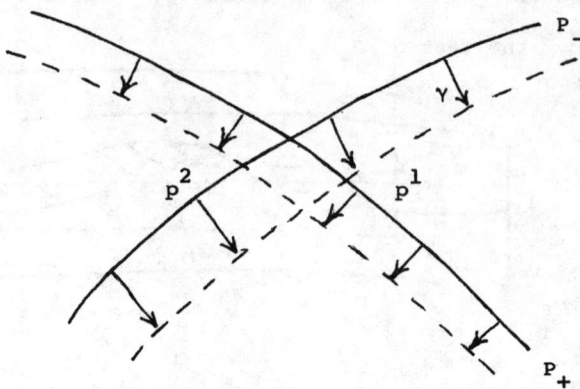

The only places where $S^n \times W$ can intersect $g'(S^n \times Y)$ are in the copies

of L above p^1 and p^2. Above p^1 we have

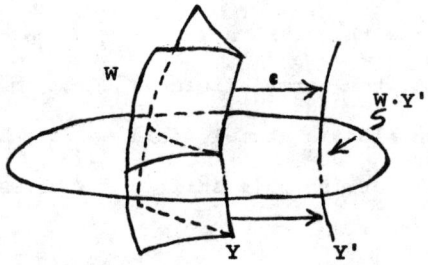

with intersection $W \cdot Y'$. Above p^2 we have

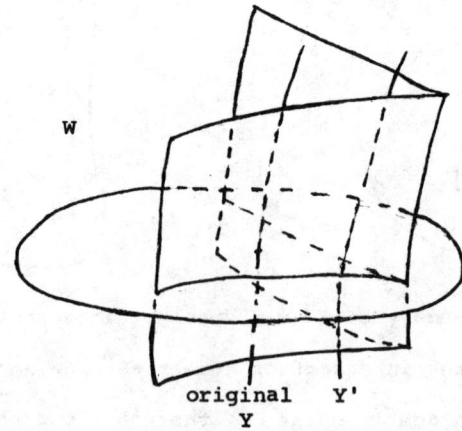

and intersection $W \cdot Y' + r(\chi(\zeta^k))$ where ζ^k is the subbundle of

$\nu(Y^k \hookrightarrow L^{2k+1})$ normal to ϵ. Let g_d be the element of Λ associated to the double point d when we order the sheets so that p_+ is first and p_- is second. The intersection associated to p^1 is $g_d \cdot (W \cdot Y') \in \Lambda$, and the intersection associated to p^2 is $(-1)^n \bar{g}_d \cdot [W \cdot Y' + r \chi(\zeta^k)]$. The total intersection is

$$\sum_{\substack{\text{double points} \\ d}} (g_d + (-1)^n \bar{g}_d)(W \cdot Y') + (-1)^n \bar{g}_d r \chi(\zeta)$$

and hence

(*) $\qquad q_{f \times 1_L}(x \otimes y) = \sum_{\substack{\text{double points} \\ d}} [\frac{1}{2r}(g_d + (-1)^n \bar{g}_d)(W \cdot Y') + \frac{1}{2}(-1)^n \bar{g}_d \chi(\zeta)]$

in $Q_{n+k+1} \otimes \mathbb{Q}/\mathbb{Z}$.

<u>Case A</u>: $2k + 1 \equiv 3(4)$.

Here k is odd and hence $\chi(\zeta^k) = 0$, and $Q_{n+k+1} = Q_n$. Thus if $\alpha \in \Lambda$, then $[\alpha] \otimes r/s = (-1)^n [\bar{\alpha}] \otimes r/s$ in $Q_{n+k+1} \otimes \mathbb{Q}/\mathbb{Z}$. Hence

$$q_{f \times 1_L}(x \otimes y) = \sum_{\substack{\text{double points} \\ d}} [\frac{1}{2r}(g_d + (-1)^n \bar{g}_d) \cdot (W \cdot Y')]$$

$$= \sum_{\substack{\text{double points} \\ d}} [\frac{1}{2}(g_d + (-1)^n \bar{g}_d) \cdot \frac{1}{r}(W \cdot Y')]$$

$$= [\sum_{\substack{\text{double points} \\ d}} \frac{1}{2}(g_d + (-1)^n \bar{g}_d)] \otimes \ell(y,y)$$

$$= [\frac{1}{2} \lambda_s(x,x)] \otimes \ell(y,y).$$

<u>Case B</u>: $2k + 1 \equiv 1(4)$.

Here k is even. In Q_{n+k+1}, $[\alpha \otimes r/s] = -[(-1)^n \bar{\alpha} \otimes r/s]$. In this case (*) becomes

$$q_{f \times 1_L}(x \otimes y) = [\sum_{\substack{\text{double points} \\ d}} (-1)^n \bar{g}_d \otimes \frac{1}{2} \chi(\zeta^k)] \ .$$

Since $\frac{1}{2}\chi(\zeta^k)$ is of order 2 in $\mathbb{Q}/\mathbb{Z}$, this sum is equal to

$$\left[\sum_{\substack{\text{double points} \\ d}} (-1)^{n+k+1} \bar{g}_d] \otimes \frac{1}{2} \chi(\zeta^k) \right]$$

$$= \mu_f(x) \otimes \frac{1}{2} \chi(\zeta^k) \in Q_{n+k+1} \otimes \mathbb{Z}/2 = Q_n \otimes \mathbb{Z}/2 \ .$$

To complete the proof of 8.3 we must show that $\frac{1}{2}\chi(\zeta^k) = \ell(y,y) \in \mathbb{Q}/\mathbb{Z}$ when k is even. But $\frac{1}{2}\chi(\zeta^k)$ is $\langle w_k(\zeta^k),[Y^k]\rangle = \langle w_k(\nu_{Y \hookrightarrow L}),[Y^k]\rangle$, and we have already seen that via Spanier-Whitehead duality the latter is $\langle v_k(\nu_K),[Y^k]\rangle = \ell(y,y) \in \mathbb{Z}/2 \hookrightarrow \mathbb{Q}/\mathbb{Z}$. This completes our local product formula.

We are now in a position to analyse the question of existence of a resolution for ℓ and $q_{f \times 1_L}$ as required by IV.3.5. First assume $\ell = 2k + 1 \equiv 3(4)$. Then, by III.3, there is a resolution

$$0 \longrightarrow A_k \longrightarrow F_k \longrightarrow \text{Tor } H_k(L) \longrightarrow 0$$

and a symmetric pairing $I: F_k \otimes F_k \to \mathbb{Q}$ lifting the linking pairing on Tor $H_k(L)$ such that $I|: A_k \otimes F_k \to \mathbb{Z}$ is non-singular. We have a resolution

$$0 \longrightarrow K_n(f) \otimes A_k \longrightarrow K_n(f) \otimes F_k \longrightarrow \text{Tor } K_{n+k}(f \times 1_L) \longrightarrow 0 .$$

<u>Claim</u>: The pairing $\lambda \otimes I: (K_n(f) \otimes F_k) \times (K_n(f) \otimes F_k) \to \Lambda \otimes \mathbb{Q}$ satisfies II.3.2, 1)-5) and

$$\text{ad}(\lambda \otimes I|): K_1(f) \otimes A_k \longrightarrow (K_n(f) \otimes F_k)^*$$

is a simple isomorphism.

<u>Proof</u>: That $\lambda \otimes I|$ satisfies II.3.2 follows immediately from the previous

calculation for ℓ and $q_f \times 1_L$. That it is non-singular is obvious.

This proves IV.1.1 in the case $\ell \equiv 3(4)$.

If $\ell = 4k + 1$ and $d(L^\ell) = 0$, let $T \hookrightarrow \text{Tor } H_{2k}(L^{2k+1})$ be a subgroup on which the linking pairing vanishes and with the following sequence being exact

$$T \xrightarrow{\ i\ } \text{Tor } H_{2k}(L^{2k+1}) \xrightarrow{\ i^*\ } T^*.$$

The nice submodule $K_n(f) \otimes T \hookrightarrow \text{Tor } H_{n+2k}(f \times 1_L)$ is a subkernel. This follows from IV.1.2 and IV.1.3. Theorem II.3.5 implies that $\sigma(f \times 1_L) = 0$.

Let X^5 be the simply connected manifold of de Rham invariant 1 as in chapter 3. Define $\varphi : L_n^s(\pi) \to L_{n+1}^s(\pi)$ (or $\varphi : L_n(\pi) \to L_{n+1}(\pi)$) by crossing any normal map with X^5. If $d(L^{4k+1}) = 1$, then $L' = L - X^5 \times \underbrace{(\mathbb{C}P^2 \times \ldots \times \mathbb{C}P^2)}_{k-1}$, has de Rham invariant equal to 0. Thus $\sigma(f \times 1_{L'}) = 0$. Hence $\sigma(f \times 1_L) = \sigma(f \times 1_{X^5 \times (\mathbb{C}P^2 \times \ldots \times \mathbb{C}P^2)})$. The latter obstruction is equal by periodicity to $\sigma(f \times 1_{X^5})$ which is $\varphi(\sigma(f))$.

This proves that in general $\sigma(f \times 1_{L^{4k+1}}) = \varphi(\sigma(f)) \cdot d(L)$.

Section IV.2 - Odd dimensional normal maps cross even dimensional manifolds. In this section we prove theorem IV.1.1 in the case
$(f : M^{2n-1} \to N^{2n-1}) \times L^{2\ell}$. We cannot arrange that f has only one non-zero kernel group. Instead we follow Wall and divide f into two normal maps with boundary each with only one non-zero kernel group. We do surgery on one of the pieces cross L until the map here is a simply homotopy equivalence of pairs. Then we consider the other piece cross L and do surgery on it relative to what we have already done on its boundary.

If $f : M^{2n-1} \to N^{2n-1}$ is a degree one normal map with $f|\partial M$ a simple homotopy equivalence, then we can do surgery on f so that $K_*(f) = 0$ for $* \leq n - 2$. Let $\{x_1, \ldots, x_\ell\} \in K_{n-1}(f)$ be a generating set. Represent these classes by disjointly embedded spheres $S_i^{n-1} \hookrightarrow M$ with trivialized

normal bundles. (The normal bundles are trivialized by the bundle data

covering f.) Let $\coprod_{i=1}^{\ell} S_i^{n-1} \times D^n$ be a tubular neighborhood of these

spheres and let $U = D^{2n-1} \cup \coprod_{i=1}^{\ell} S_i^{n-1} \times D^n$ as in the diagram below

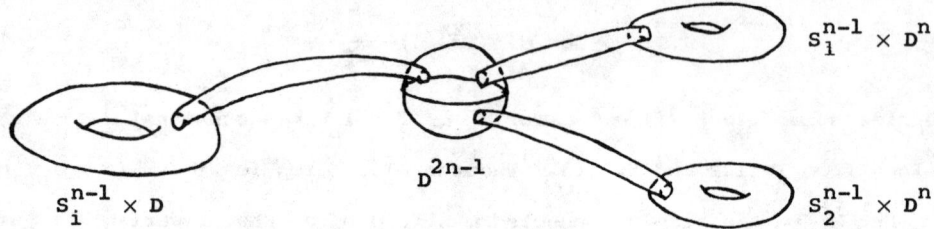

We can assume f: $U \to D^{2n-1}$, and that f| (M−U): $M - U \to N - D^{2n-1}$. Let

$M - U$ be M_0 and $N - D^{2n-1}$ be N_0. The fact that ∂U has the structure of

$\#_{i=1}^{\ell} S_i^{n-1} \times S^{n-1}$ tells us that $K_*(f|\partial U)$ is 0 for $* \neq n - 1$ and that

$(K_{n-1}(f|\partial U), \lambda, \mu)$ is isomorphic to the hyperbolic form $\bigoplus_{i=1}^{\ell} \Lambda(e_i, f_i)$, with

$\lambda(e_i, e_j) = \lambda(f_i, f_j) = 0$, $\lambda(e_i, f_j) = \delta_{ij}$, and $\mu(e_i) = \mu(f_i) = 0$. The

element e_i is the class of $S_i \times \{pt\}$ and f_i is the class of the dual

sphere. The kernel sequence for the pair $(M_0, \partial U)$ is trivial except for

$$0 \longrightarrow K_n(M_0, \partial U) \longrightarrow K_{n-1}(\partial U) \longrightarrow K_{n-1}(M_0) \longrightarrow 0.$$

We can assume that $K_n(M_0, \partial U)$ is free. It provides a based subkernel in

the hyperbolic form $K_{n-1}(\partial U)$. The class of this subkernel determines the

surgery obstruction of f in $L_{2n-1}^s(\pi_1(N))$, see [15] page 56. We denote

the above sequence of free based kernel groups

$$0 \longrightarrow K_R \longrightarrow K_\partial \longrightarrow K_A \longrightarrow 0.$$

The pairing $K_R \times K_A \xrightarrow{\lambda} \Lambda$ is non-singular. (In fact the adjoint of λ is

a simple isomorphism.)

     Our method of proof is to assume that the index of L is 0 and

then do surgery on $f \times 1_L : M_0 \times L \to N_0 \times L$ to produce a simple homotopy

equivalence of pairs.

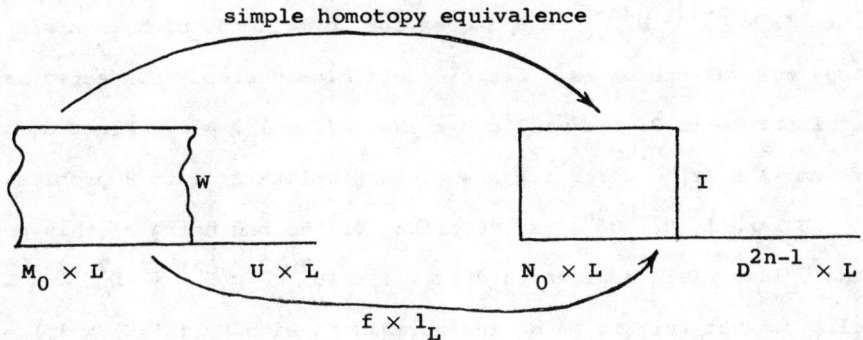

After having done this we will have to do surgery on $W \cup U \times L \to D^{2n-1} \times L$

relative to its boundary. Since this is a normal map between simply

connected odd dimensional manifolds, we can perform the surgery and re-

place the map by a homotopy equivalence.

This will prove that when $I(L^{2\ell}) = 0$ then $\sigma(f \times 1_L) = 0$. The com-

plete product formula, $\sigma(f \times 1_L) = \sigma(f) \cdot I(L)$, then follows easily from

this and the fact that $\sigma(f \times 1_{\mathbb{C}P^{2n}}) = \sigma(f)$. To prove the product formula

in $L^h(\pi)$ simply delete all reference to the based structure.

<u>Proposition</u> IV.2.1: Suppose f: $P^{2m} \to Q^{2m}$ is an s-nice normal map and

μ: Tor $K_m(f) \to Q_m$ vanishes. Then given a subkernel $S \subset K_m(f)/\text{Tor}$ there is

a normal bordism F: $W \to Q \times I$ from f to a simple homotopy equivalence

with

1) $K_*(W)$ and $K_*(W, \partial)$ nice, based Λ — modules,

2) $K_*(W, P) = 0$ for $* \geq m + 2$,

3) $K_*(W, P \to K_{*-1}(P)$ a based isomorphism for $* \leq m$, and with

4) $K_{m+1}(W, P) \overset{\partial}{\to} K_m(P)$ a based isomorphism onto $S \subset K_m(P)/\text{Tor}$.

<u>Proof</u>: The low dimensional surgery as described in I.3 produces a bordism

from f to f' with $K_n(f') = K_n(f)/\text{Tor} \oplus \Lambda^r \oplus \Lambda^r$. Surgery along a basis

for $S \subset K_n(f)/\text{Tor}$ and a basis for the second Λ^r factor as described in the

note following II.1.4 then produce the required normal bordism from f to

a simple homotopy equivalence.

Let $f: M^{2n-1} \to N^{2n-1}$ be a degree one normal map which is a simple

homotopy equivalence on ∂M. Let $L^{2\ell}$ be a closed simply connected manifold

whose signature is 0. We will prove that $\sigma(f \times 1_L) = 0$. First do surgery

on f until $K_i(f) = 0$ for $i \leq n - 2$. Now split f into 2 problems

$f: M_0 \to N_0$ and $f: U \to D^{2n-1}$ as described at the beginning of this section.

The analysis of IV.1 applies to $f|\partial U \times 1_L: \partial U^{2n-2} \times L^{2\ell} \to \partial D^{2n-2} \times L^{2\ell}$.

It tells us that this is an s-nice normal map with $K_i((f|\partial U) \times 1_L) =$

$K_\partial \otimes H_{i-n+1}(L)$. The linking and intersection pairings on these kernel

groups are the tensor product of the pairings on $H_\ell(L)$ with the inter-

section pairing on K_∂. The quadratic form vanishes on

Tor $K_{n+\ell-1}(f|\partial U \times 1_L)$. Let $S_L \hookrightarrow H_\ell(L^{2\ell})/\mathrm{Tor}$ be a subkernel for the inter-

section form. $K_\partial \otimes S_L$ is a subkernel of $K_{n+\ell-1}(f|\partial U \times 1_L)/\mathrm{Tor}$. Let

$W^{2n+2\ell} \xrightarrow{F} \partial D^{2n-1} \times L^{2\ell} \times I$ be the normal bordism of $f|\partial U \times 1_L$ to a simple

homotopy equivalence as in IV.2.1. Form $V = M_0 \times L \times I \underset{\substack{\text{collar} \\ \text{on } \partial U \times L}}{\cup} W \times I \to$

$(N_0 \times L) \times I$.

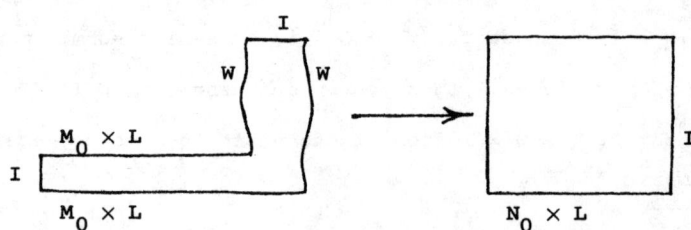

This is a normal bordism from $f \times 1_L$ to a normal map $g: X^{2n+2\ell-1} \to N_0 \times L$,

$(X = M_0 \times L \cup W)$. The map g restricted to ∂X is a simple homotopy

equivalence. We have the exact sequence of the pair $(M_0 \times L \cup W, M_0 \times L)$.

$$\cdots \longrightarrow K_*(W, \partial U \times L) \longrightarrow K_*(M_0 \times L) \longrightarrow K_*(g) \longrightarrow K_*(W, \partial U \times L) \longrightarrow K_{*-1}(M_0 \times L).$$

This implies that

$$K_*(g) = \begin{cases} K_R \otimes H_{*-n}(L) & * \leq n + \ell - 2 \\ \\ K_A \otimes H_{*-n+1}(L) & * \geq n + \ell + 1. \end{cases}$$

On these modules the intersection and linking pairings are the tensor product of $\lambda: K_R \times K_A \to \Lambda$ with the pairings on $H_*(L)$. Also near the middle dimension we have short exact sequences

$$0 \longrightarrow K_A \otimes H_{\ell+1}(L) \longrightarrow K_{n+\ell}(g) \longrightarrow K_R \otimes S \longrightarrow 0$$

$$0 \longrightarrow K_A \otimes H_{\ell}(L)/S \longrightarrow K_{n+\ell-1}(g) \longrightarrow K_R \otimes H_{\ell-1}(L) \longrightarrow 0$$

In both cases the submodule $K_A \otimes H_*(L)$ is represented by product cycles in $M_0 \times L$. Both sequences are split. Cycles representing $K_R \otimes S \subset K_{n+\ell}(g)$ and $K_R \otimes H_{\ell-1}(L) \subset K_{n+\ell-1}(g)$ are of the form

$$X \times Z \cup P \hookrightarrow M_0 \times L \cup W$$

where X is a relative cycle in $(M_0, \partial U)$; Z is a cycle in L; and P is a relative cycle in W whose boundary is $\partial X \times Z$

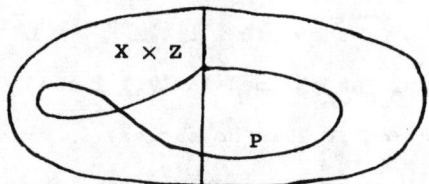

From these descriptions it follows easily that g is an s-nice normal map. Hence $\sigma(g) \in L^s_{2n+2\ell-1}(\pi_1(N))$ is determined by Tor $K_{n+\ell-1}(g)$ together with its linking pairing and quadratic refinement. We have a short exact sequence

$$(*) \quad 0 \longrightarrow K_A \otimes \text{Tor } H_{\ell}(L) \longrightarrow \text{Tor } K_{n+\ell-1}(g) \longrightarrow K_R \otimes \text{Tor } H_{\ell-1}(L) \longrightarrow 0.$$

The linking pairing and quadratic refinement vanish on $K_A \otimes \text{Tor } H_{\ell}(L)$. The reason is that $\alpha \otimes t \in K_A \otimes \text{Tor } H_{\ell}(L^{2\ell})$ is represented by a product cycle in $M_0 \times L \subset M_0 \times L \cup W$. If $S^{n-1} \hookrightarrow M_0$ represents α and $Y^{\ell} \hookrightarrow L^{2\ell}$ represents t, then $S^{n-1} \times Y \hookrightarrow M_0 \times L$ represents $\alpha \otimes t$. Any normal field for Y^{ℓ} in $L^{2\ell}$ is an appropriate normal field to calculate $q(\alpha \otimes t)$ since it will extend over $D^n \times Y$. (Such a normal field exists since $Y^{\ell} \cdot Y^{\ell} = 0$.)

With these choices it is clear that $q(\alpha \otimes t) = 0$. Thus the exact sequence
(*) is of the form

$$T \xrightarrow{\ i\ } \text{Tor } K_{n+\ell-1}(g) \xrightarrow{\ i*\ } T^*,$$

with $q|T = 0$. Theorem 6.5 implies that $\sigma(g) = 0$.

We do surgery on g relative to $g|\partial(M_0 \times L \cup W)$ to make g a
simple homotopy equivalence of pairs.

As we remarked earlier surgery on the "other side" relative to its
boundary, $W \cup U \times L \to D^{2n-1} \times L$, is always possible. In the end we have
constructed a normal bordism for $f \times 1_L : M \times L \to N \times L$ to a simple homotopy equivalence. Consquently, $\sigma(f \times 1_{L^{2\ell}}) = 0$ if $I(L^{2\ell}) = 0$.

Section IV.3 - Odd dimensional normal map cross odd dimensional manifolds.

In this section we consider the last case of the product formula. We
show that if $d(L^{2\ell+1}) = 0$ then $\sigma(f \times 1_L : M^{2n-1} \times L^{2\ell+1} \to N^{2n-1} \times L^{2\ell+1})$
$= 0$. The general formula as claimed in IV.1.1 then follows easily by
additivity. As in section IV.2 we do surgery in $f: M^{2n-1} \to N^{2n-1}$ until
$K_i(f) = 0$ for $i < n - 1$, then we split f into $f: M_0 \to N_0$ and
$f: U \to D^{2n-1}$. First we show that if $d(L^{2\ell+1}) = 0$, then we can assume
that $\text{Tor } H_\ell(L)$ has a subkernel. This then provides a subkernel for
$f|\partial U \times L$.

Lemma IV.3.1: Let $L^{2\ell+1}$ be a closed oriented manifold.

   a) If $2\ell + 1 \equiv 1(4)$ and $d(L) = 0$, then there is a subgroup
       $T \subset \text{Tor } H_\ell(L)$ on which the linking pairing vanishes and so that
       the following sequence is exact.

$$T \xrightarrow{\ i\ } \text{Tor } H_\ell(L) \xrightarrow{\ i*\ } T^*.$$

   b) If $2\ell + 1 \equiv 3(4)$, then L is bordant to a manifold L' so that
       there is a subgroup $T \subset \text{Tor } H_\ell(L')$ as in a).

<u>Proof</u>: a) follows immediately from III.4. In case b) we know that there is a short exact sequence

$$0 \longrightarrow A \longrightarrow F \longrightarrow \text{Tor } H_\ell(L) \longrightarrow 0$$

and a symmetric pairing

$$I: F \otimes F \longrightarrow \mathbb{Q}$$

which refines the linking pairing. In addition $I: A \otimes F \to \mathbb{Z}$ is non-singular. We use the matrix $I: A \otimes A \to \mathbb{Z}$ to build a plumbing diagram of 2-spheres. The resulting 4 dimensional manifold, W, has a boundary ∂W with $H_1(\partial W) = \text{Tor } H_\ell(L)$ and with the negative linking pairing. Take $\partial W \times \mathbb{CP}^{2r}$ and do two dimensional surgery to make it simply connected. (Here $4r + 3 = 2\ell + 1$.) Call the result $X^{2\ell+1}$. Let $L' = X \# L$. Since X bounds, L and L' are bordant. The linking pairings on $\text{Tor } H_\ell(L')$ has a decomposition

$$(\text{Tor } H_\ell(L'), \ell) \cong (\text{Tor } H_\ell(L), \ell) \oplus (\text{Tor } H_\ell(L), -\ell).$$

The diagonal copy of $\text{Tor } H_\ell(L')$ provides the required subspace T. Since L and L' are bordant $\sigma(f \times 1_L) = \sigma(f \times 1_{L'})$. Thus for calculating surgery obstructions we can always assume that if $d(L^{2\ell+1}) = 0$, then such $T \subset \text{Tor } H_\ell(L)$ exist.

Let $f: M^{2n-1} \to N^{2n-1}$ is a degree 1 normal map. As in section IV.2, we do surgery on f to make $K_i(f) = 0$ for $i \leq n - 2$, and let $f_0: M_0^{2n-1} \to N_0$, $f| : U \to D^{2n-1}$ be as before. We will do surgery on

$$f| \times 1_L: \partial U \times L^{2\ell+1} \longrightarrow S^{2n-1} \times L^{2\ell+1}$$

to make this normal map a simple homotopy equivalence (assuming $d(L^{2\ell+1}) = 0$). This surgery will be well adapted to this subkernel. To prove such surgery is possible requires some analysis of immersions of m-dimensional $\mathbb{Z}/k$-manifolds in $2m-1$ dimensional manifolds.

Let $f: P^{2m-1} \to R^{2m-1}$ be an s-nice degree 1 normal map and let

$\tilde{f}\colon \nu_p \to \xi$ be the bundle map covering f. Suppose $T \subset \mathrm{Tor}\ K_{m-1}(f)$ is a subkernel with generating set $\{t_1,\ldots,t_r\}$ with the order of t_i being n_i. There are relative $\mathbb{Z}/n_i$-bordism elements $(X_i^m,Y_i^{m+1}) \overset{\varphi}{\to} (P,R)$, representing those t_i whose order is a power of 2.

<u>Lemma</u> IV.3.2: The bundle data can be used to immerse X_i^m into P so that $\delta X_i \cap (X_i - \delta X_i)$ is empty.

<u>Proof</u>: Reduce $\nu_{\delta Y_i} - \varphi^*\xi$ to an $(n-1)$ plane bundle over δY_i^m. This induces an embedding $\delta X_i^{m-1} \hookrightarrow P^{2m-1}$ with a normal field $\boldsymbol{\epsilon}$. Arrange that the n_i sheets of X come off δX_i along $-\boldsymbol{\epsilon}$. Let $\delta X_i'$ be a copy of δX_i pushed off along $+\boldsymbol{\epsilon}$. Since $q([\delta X_i']) = 0$, we have $\frac{1}{2n_i}(\delta X_i' \cdot X_i) = 0$ in $\mathbb{Q}/\mathbb{Z} \otimes \mathbb{Q}_m$. Thus we can deform δX_i be a regular homotopy (which intersects δX_i) to change $\delta X_i' \cdot X_i$ to 0. (See II.3.3.) Now we extend the embedding of δX_i to an immersion of X_i^m into P^{2m-1}. To do this consider $\bar{Y}_i$, which is Y_i cut open along δY_i, and $\bar{X}_i$ which is X_i cut open along δX_i. The boundary of $\bar{Y}_i$, $\partial \bar{Y}_i$, is $\bar{X}_i \cup \coprod_{j=1}^{n_i} \delta Y_i$. We have a reduction of $\nu_{Y_i} - \varphi^*\xi$ to an $(m-1)$ plane bundle over $\coprod_{j=1}^{n_i} \delta Y_i$. To extend the embedding of $\delta X_i \hookrightarrow P$ to an immersion of X_i, we must extend this bundle reduction over all of $\bar{Y}_i$. The obstructions to doing this lie in $H^*(\bar{Y}_i, \coprod_{j=1}^{n_i} \delta Y_i;\ \pi_*(SO(N)/SO(m-1)))$. If $\bar{Y}$ and $\bar{X}$ are connected and simply connected (as we can and do assume), then all these cohomology groups vanish.

The resulting immersion of X_i^m in P^{2m-1} has 0 algebraic intersections with $\delta X_i'$. Thus be deforming it we can make $\delta X_i' \cap X_i = \emptyset$. If we have such an immersion $X_i^m \looparrowright P^{2m-1}$ with $(X_i - \delta X_i) \cap \delta X_i = \emptyset$ then the self-intersection of the immersion is a union of circles whose preimages in X_i miss δX_i. There are two types of circles—those whose preimages are two circles and those whose preimage is one circle. We call components of the second type doubly covered circles of self-intersection. Any such circle must be of order two in $\pi_1(P^{2m-1})$ and its effect on the orientation must be $(-1)^{m-1}$.

Define $\mu(X_i)$ to be the sum, in Q_m, of the classes determined by the doubly covered circles of self-intersection. Notice that each circle represents an element of order 2 in Q_m.

Theorem IV.3.3: a) If n_i is even then $\mu(X_i)$ depends only on
$[X_i] \in K_m(f; (\mathbb{Z}/n_i)[\pi_1(R)])$, not on the particular immersion chosen. We can deform X_i to remove all the doubly covered circles of self-intersection if $\mu(X_i) = 0$.

b) If n_i is odd and we have an immersed $\mathbb{Z}/n_i$-manifold X_i as above then we can deform X_i by regular homotopy to remove all the double covered circles of self-intersection.

Proof: First suppose for n_i either even or odd we have made $\mu(X_i) = 0$. This means that the doubly covered circles of self-intersection $\{s_i^1\}_{i=1}^{t}$ can be paired up $(S_1,S_2),(S_3,S_4),\ldots,(S_{t-1},S_t)$ so that the two circles in each pair represent the same element in $\pi_1(P)$. The following construction allows us to cancel each of these pairs. Pick points $P_{2i-1} \in S_{2i-1}$ and $P_{2i} \in S_{2i}$. Order the two sheets near each of these points and connect P_{2i-1} to P_{2i} by an arc on each sheet, A and B.

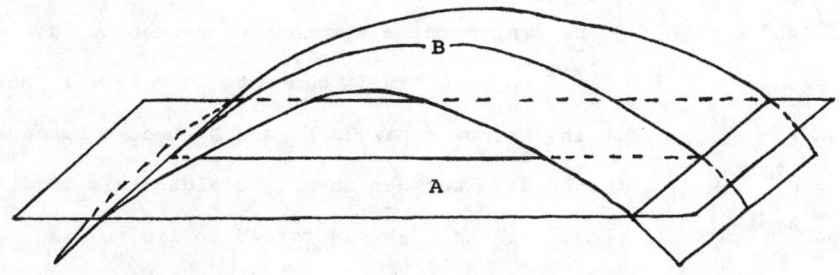

The loop A_*B^{-1} bounds a 2 disk in P since S_{2i-1} and S_{2i} represent the same element in $\pi_1(P)$. Pushing a neighborhood of B across the disk past A changes the self-intersection by replacing S_{2i-1} and S_{2i} by a single circle of self-intersection whose preimage in X is two circles.

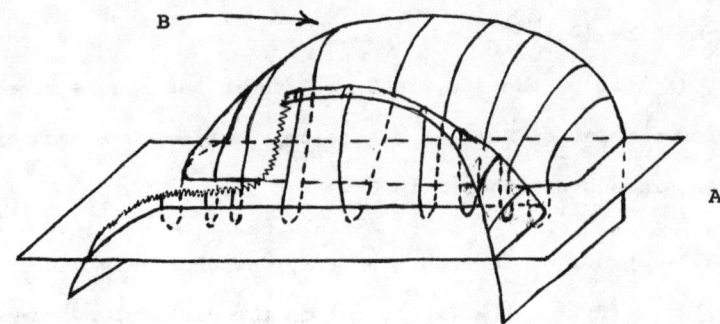

There is another description of μ which is useful in proving a)
and b). Consider the immersion of X_i in P as an immersion into
$P \times \{0\} \hookrightarrow P \times I$. In $P \times I$ deform X_i by a regular homotopy <u>relative to</u>
δX_i until it is in general position. Take its geometric self-intersection
in the usual sense in Q_m. This is also $\mu(X_i)$. The point is that circles
of self-intersection whose preimages are two circles can be pulled apart
in $P \times I$ by simply shifting the map in the I factor near one of the components of the preimage. For doubly covered circles this process does
not work, and in fact when we put the immersion in general position we are
left with one point of self-intersection for each doubly covered circle.
The element in Q_m associated to such a point is the group element represented by the original circle in $\pi_1(P)$.

Suppose $\alpha \in K_m(f; (\mathbb{Z}/n_i)[\pi_1(R)])$ for n_i even, and suppose
$X^m \leadsto P^{2m-1}$ and $Z^m \leadsto P^{2m-1}$ are $\mathbb{Z}/n_i$ bordism elements representing α. We
do surgery on $f: P^{2m-1} \to R^{2m-1}$ to make $f(m-1)$ connected. This will not
change $\mu(X)$ or $\mu(Z)$, but in the new manifold X and Z become bordant.
Let $W^{m+1} \to P^{2m-1} \times I$ be the bordism between them. Consider this bordism
as one in $P^{2m-1} \times I \times I$ whose Bochstein δW is forced to lie in $P \times \{0\}$
$\times I$. The argument in the proof of proposition II.1.1 shows that we can
immerse W into $P \times I \times I$ connecting the immersions of X and Z in $P \times I$.
δW becomes an immersed bordism in $P \times \{0\} \times I$ connecting
$\delta X \hookrightarrow P \times \{0\} \times \{0\}$ with $\delta Z \hookrightarrow P \times \{0\} \times \{1\}$. The difference

$\delta Z' \cdot Z - \delta X' \cdot X = \alpha + (-1)^m \bar{\alpha}$ where $\alpha \in Q_m$ is the self-intersection of δW_i in $P \times I$. Since both terms on the left are 0 it follows that $\alpha + (-1)^m \bar{\alpha} = 0$. Now we are in a position to compare $\mu(X)$ and $\mu(Z)$. We break the argument into 2 cases.

I: If δW is embedded in $P \times \{0\} \times I$, then $\mu(X) = \mu(Z)$.

<u>Proof</u>: The self-intersection of W is a one manifold whose boundary is the self-intersection of Z in $P \times I$ minus the self-intersection of X in $P \times I$. Thus $\mu(Z) - \mu(X) = 0$.

II. $\delta W = \delta X \times I$ immersed in $P \times \{0\} \times I$. Then $\mu(Z) = \mu(X) + n_i^2 \mu(\delta W)$.

<u>Proof</u>: In light of case I, $\mu(Z)$ agrees with the self-intersection of the following manifold

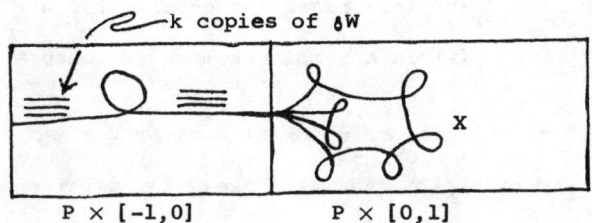

Clearly, this self-intersection is $\mu(X) + n_i^2 \mu(\delta W)$.

Since $\mu(\delta W)$ must be of order 2 in Q_m, if n_i is even the $\mu(X) = \mu(Z)$. This proves IV.3.3, a. If n_i is odd, then $\mu(Z) = \mu(X) + \mu(\delta W)$. To complete the proof of IV.3.3,b, it suffices to show that given a nice immersion $X \rightsquigarrow P^{2m-1}$ that there is a regular homotopy of it to another nice immersion of X into P^{2m-1} so that the restriction of the regular homotopy to δX given an immersion $\delta X \times I \rightsquigarrow P^{2m-1} \times I$ with self-intersection any prescribed $\alpha \in Q_m$ with the property that $\alpha + (-1)^m \alpha = 0$. To do this we take a regular homotopy of $\delta X \times I \rightsquigarrow P^{2m-1} \times I$ with self-intersection the preassigned $\alpha \in Q_m$. By homotopy extension, we extend this to a regular homotopy of X. If $\alpha + (-1)^m \bar{\alpha} = 0$ in Λ, then the new

immersion will have the property that $\partial X' \cdot X = 0$. Thus there is a further
deformation of it relative to ∂X which produces an immersion with
$\partial X \cap (X - \partial X) = \emptyset$.

The next proposition is the key one of doing the correct type of
surgery in $f|\ \partial U \times L^{2\ell+1} \to \partial D \times L^{2\ell+1}$.

<u>Proposition IV.3.4</u>: Let $f: P^{2m-1} \to Q^{2m-1}$ be an s-nice normal map. Sup-
pose that $T \subset \text{Tor } K_{m-1}(f)$ is a subkernel with natural generating set
$\{t_1,\ldots,t_r\}$. Suppose further that for each t_i whose order is even we
have a nice immersion $X_i^m \looparrowright P^{2m-1}$ with ∂X_i representing t_i and $\mu(X_i) = 0$.
Then there is a normal bordism $F: W \to R \times I$ from F to a simple homotopy
equivalence such that:

1) $K_*(W)$ and $K_*(W,P)$ are based, nice Λ-modules,

2) $K_*(W,P) = 0$ for $* \geq m + 1$,

3) $K_*(W,P) \overset{\partial}{\to} K_{*-1}(P)$ is a based isomorphism for $* \leq m - 1$, and

4) $K_m(W,P) \overset{\partial}{\to} K_{m-1}(P)$ is a simple isomorphism onto $K_{m-1}(P)/\text{Tor} \oplus T$.

<u>Proof</u>: By I.3.5 we can produce a normal bordism $V \to R \times I$ from f to
$f': P' \to R$ satisfying 1), 2), and 3) above. In addition $K_m(V,P) \overset{\partial}{\to} K_{m-1}(P)$
is a simple isomorphism onto $K_{m-1}(P)/\text{Tor}$. The kernel groups $K_*(f')$ all
vanish except for $K_{m-1}(f')$ which is isomorphic to $\text{Tor } K_{m-1}(f)$. To complete the proposition we must construct a normal bordism from f' to a
simple homotopy equivalence, $G: V' \to R \times I$ so that $K_*(V',P')$ is 0 for all
$* \neq m$, and $K_m(V',P') \overset{\partial}{\to} K_{m-1}(P')$ is a simple isomorphism onto T.

Let $A \to F \to T$ be a based resolution for T with the basis for F
being $(z_1,\ldots,z_r)$. Represent the z_i by disjointly embedded spheres
$S_i^{m-1} \hookrightarrow P'$. The normal bundles of these spheres are trivialized by the
bundle data covering f'. The classes $n_i[S_i^{m-1}]$ are zero in $K_{m-1}(f')$. Thus
there are immersed manifolds $X_i^m \looparrowright P'$ with $\partial X_i = n_i S_i$. By the Hurewicz
theorem we can take each X_i to be a sphere with holes. Each of these is
immersed by the bundle data covering the normal map. The normal bundles

are thus trivialized by a trivialization extending the ones for the S_i.

By hypothesis we can choose the X_i so that $\mu[X_i] = 0$. Since the linking

pairing and its quadratic refinement vanish on T, we can deform the X_i

by regular homotopy (moving the S_i) until $\delta X_i \cdot X_j = 0$ for $i \neq j$ and

$\delta X_i' \cdot X_i = 0$ (see II.3.3). Once this is accomplished we can deform the X_i

relative to their boundaries so that $\delta X_i \cap X_j = \emptyset$ for $i \neq j$ and

$\delta X_i \cap (X_i - \delta X_i) = \emptyset$. Since $\mu([X_i]) = 0$, we can in addition suppose that

the self-intersections of each X_i are circles whose preimages in $X_i - \delta X_i$

are two circles.

It is time now to remove the remaining self-intersections and inter-

sections of the X_i. For this we need embedded spheres with trivial normal

bundles $S_i^* \hookrightarrow P'$ so that

$$S_i^* \cap X_j = \begin{cases} \emptyset & i \neq j \\ \\ 1 \text{ pt} & i = j \end{cases} .$$

These spheres represent elements in $K_{m-1}(f')$ which project via $i*$ to the

canonical generating set for $T*$. Using an argument as in II.3.3 we can

arrange the desired intersections. Consider now a circle of intersection

of X_i with X_j. We deform the intersection in X_j until it bounds a small

2 disk in the D^m normal to S_j^* at the point $S_j^* \cap X_j$. We can assume that

a neighborhood of the circle of intersection in X_i is $D^{m-1} \times S^1 \hookrightarrow S_j^* \times D^m$

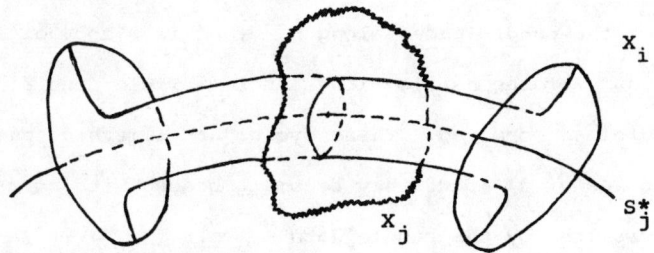

Replace $D^{m-1} \times S^1$ in X_i by $(S_j^* - D^{m-1}) \times S^1$

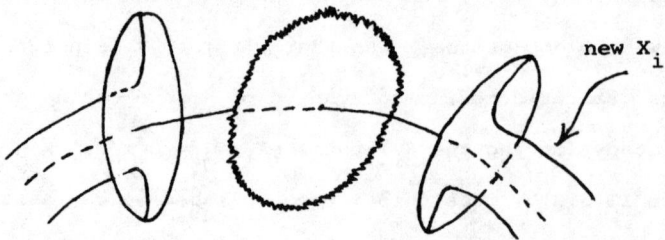

new X_i

This changes the homotopy class of X_i but not its homology class or the fact that it is immersed with trivial normal bundle compatibly with the bundle data. It also removes one circle of intersection.

The argument to remove the circles of self-intersection is exactly the same (once we know that their preimages are two circles not one). In the end we have the X_i disjointly embedded with trivial normal bundle compatibly with the bundle data.

Do surgery on the $S_i^{m-1} \hookrightarrow P'^{2m-1}$. Let $W \xrightarrow{F} R \times I$ be the resulting normal bordism, and $f'': P'' \to R$ be the "other end". By II.3.5

$$K_*(f'') = \begin{cases} 0 & * \neq m-1, m \\ A* & * = m-1 \\ A & * = m. \end{cases}$$

The basis for $A = K_m(f'')$ is represented by

$$\{-X_1' \cup n_1 d_1', \ldots, -X_t' \cup n_t d_t'\}$$

where d_j is the case of the handle added along S_j^{m-1}; d_j' is a copy of d_j pushed out to the boundary of the handle along a normal field, and X_j' is X_j with a neighborhood of δX_j removed. These cycles are immersed spheres. Once we push the n_j copies of d_j' apart they become embedded with normal bundles which are trivialized by the bundle data.

Now do surgery on these spheres. Note that these spheres are m dimensional in P^{2m-1}. Nonetheless, we have managed to find them disjointly embedded with trivial normal bundles. Since they form a basis for $K_m(f'')$ surgery on these provides a normal bordism from f'' to a simple homotopy

equivalence. The union of this with the previous normal bordism from f
to f" is the normal bordism required in IV.3.4.

IV.3.5: Suppose, under the hypothesis of 10.4, that we have embedded
cycles $Z_1^{m-1}, \ldots, Z_r^{m-1}$ in P which represent those elements in the generating set which are of order a power of 2. If each Z_i is equipped with a
normal field e by the bundle data so that $n_i Z_i = \partial C_i^m$ with $C_i \cdot Z_j' = 0$ for
all i and j, then in the bordism W constructed in IV.3.4 the Z_i bound
disjointly embedded manifolds Y_i^m. The embeddings of the Y_i^m are compatible
(relative to the Z_i) with the bundle data of the normal map.

Proof of IV.3.5: Given a set of disjointly embedded cycles with normal
fields $Z_i^{m-1} \hookrightarrow P^{2m-1}$ which represent torsion classes in $K_{m-1}(P)$ there is a
well defined chain intersection matrix. Namely let $n_i Z_i = C_i^m$ where the n_i
sheets of C_i leave Z_i along the negative direction of the normal field
form $C_i \cdot Z_i$ for $i \neq j$ and $C_i \cdot Z_j'$. This matrix of intersections depends only
on the position of the Z_i and their normal fields not on the choice of the
C_i. Once we have done surgery to make f: $P^{2m-1} \to R^{2m-1}$ highly connected
the Z_i will be bordant, in P × I, to spheres. Use the bundle data to
immersion the bordisms relative to the Z_i in P × {0}. Pipe all the intersection and self-intersection points off the P × {1} end. This will give
up spheres $S_i^{m-1} \hookrightarrow P^{2m-1}$. The chain intersection pairing that they
generate will agree with the one for the Z_i. Thus if the Z_i generate the
0 intersection pairing so will the S_i^{m-1}. If the Z_i are a partial generating set for T we complete the set of spheres in P × {1} to a full set
keeping the chain intersection pairing trivial. Once we have such spheres
we might have to shift the ones of odd order before doing surgery as
required in IV.3.4, but we can leave the ones of even order fixed. Thus
the S_i^{m-1} which are bordant in M × I to the Z_i^{m-1} bound the cores of the
handles added in doing the surgery.

Let f be a normal map f: $(P^{2m}, \partial P) \to (R, \partial R)$ and α an element in $K_m(P, \partial P)$

which is represented by a relative bordism element

$$
\begin{array}{ccc}
(X^m, \partial X^m) & \xrightarrow{\;\Phi\;} & (P, \partial P) \\
\big\downarrow & & \big\downarrow f \\
(Y, \partial_+ Y) & \xrightarrow{\;\psi\;} & (R, \partial R)
\end{array}
$$

with $\partial Y = \partial_+ Y \cup_{\partial X} X$. Suppose that the bundle map covering f is $\tilde{f} \colon \nu_p \to \xi$. Then we can reduce $\nu_Y - \psi^* \xi$ to an m-plane bundle over Y and reduce $(\nu_Y - \psi^* \xi) \mid (\partial Y - X)$ to an (m-1) plane bundle. Restricting to X this gives an immersion $X^m \looparrowright P^{2m}$ and a normal field for $\partial X \hookrightarrow \partial P$. (Of course, generically the immersion of ∂X into ∂P is an embedding.) The normal field for $\partial X \hookrightarrow \partial P$ automatically extends over the immersion $X \looparrowright P$. We call such an immersion which when restricted to the boundary is an embedding with a normal field <u>compatible with the bundle data</u>. It is in this sense that we say that $Z_i^{m-1} \hookrightarrow P^{2m-1}$ bound disjointly embedded manifolds $Y_i^m \hookrightarrow W^{2m}$ with the embedding compatible with the bundle data.

If we have two normal maps $f_i \colon (P_i^{2m}, \partial P_i^{2n-1}) \to (R_i, \partial R_i)$ $i = 1,2$ which agree on the boundary, then we can glue them together to form a normal map $f_1 \cup f_2 \colon P_1 \cup P_2 \to R_1 \cup R_2$. Suppose we have elments $\alpha_i \in K_m(f_i, f_i \mid \partial P_i)$ which are represented by immersed relative bordism elements $(W_i^m, \partial W_i^m)$ $(P_i, \partial P_i)$ compatibly with the bundle data, and that the embeddings $\partial N_i^m \hookrightarrow \partial P_i$ are the same and have the same normal field. The union $W_1 \cup - W_2 \looparrowright P_1 \cup P_2$ is immersed by the bundle data and thus is correct for calculating the self-intersection of the class it represents.

Now we are ready to construct a normal bordism from $f \times 1_L \colon \partial U \times L^{2\ell+1} \to \partial D \times L$ to a simple homotopy equivalence (when $d(L^{2\ell+1}) = 0$). By IV.3.1 we can assume that there is a subgroup $T \subset \operatorname{Tor} H_\ell(L^{2\ell+1})$ so that

$$
T \xrightarrow{\;i\;} \operatorname{Tor} H_\ell(L) \xrightarrow{\;i^*\;} T^*
$$

is exact. The map $f \times 1_L: \partial U \times L \to \partial D \times L$ is an s-nice normal map. Its kernel groups are $K_\partial \underset{\mathbb{Z}}{\otimes} H_*(L)$, and the pairings are the tensor product of the intersection pairing on K_∂ and the usual linking and intersection pairings on $H_*(L)$. Thus $K_\partial \otimes T \hookrightarrow \text{Tor } K_{n-1+\ell}(f \times 1_L)$ is a submodule on which the linking pairing vanishes and the following sequence is exact and has 0 Whitehead torsion:

$$K_\partial \otimes T \xrightarrow{\ i\ } \text{Tor } K_{n+\ell-1}(f \times 1_L) \xrightarrow{\ i*\ } K_\partial \otimes T*.$$

By IV.1.3 and IV.1.3 the quadratic refinement of the linking pairing also vanishes on $K_\partial \otimes T$. To be able to invoke IV.3.4 there is one more condition we must check, namely that there is a lifting of $K_\partial \otimes T$ into $K_{n+\ell}(f \times 1_L; \mathbb{Q}/\mathbb{Z})$ which is in the kernel of the self-intersection form. This is proved using the following lemma.

<u>Lemma</u> IV.3.6: Let $X^{\ell+1} \overset{\psi}{\to} L^{2\ell+1}$ be a $\mathbb{Z}/k$ manifold mapping into $L^{2\ell+1}$. Suppose $(\partial X) \cdot X \equiv 0(k)$. Then there is an immersion of X in L homotopic to the original map. This immersion is an embedding on ∂X and has only circles of self-intersection whose preimages in $X - \partial X$ are two circles.

<u>Proof</u>: Shift the map on ∂X to an embedding and make the k sheets of X come off of ∂X in some direction, $(-\epsilon)$. Let ζ^ℓ be a complement to ϵ in $\nu_{\partial X \subset L}$. Let $\partial X'$ be a copy of ∂X pushed out to $\partial \nu_{\partial X \subset L}$ along $-\epsilon$. Let S^ℓ denote a fiber of $\nu_{\partial X \subset L}$. The cycle $k \partial X' + (\partial X \cdot X) S^\ell$ in $\partial(\nu_{\partial X \subset L})$ bounds in $L - \text{int } \nu_{\partial X \subset L}$. Thus its intersection with itself is 0. If $\ell \equiv 0(2)$, this tells us that

$$0 = [k \partial X' + (\partial X \cdot X) S^\ell] \cdot [k \partial X' + (\partial X \cdot X) S^\ell] = k^2 \chi(\zeta) + 2k(\partial X \cdot X).$$

Since $(\partial X \cdot X) \equiv 0(k)$ it follows that $\chi(\zeta) \equiv 0(2)$. By changing the class of the section ϵ we can change $\chi(\zeta)$ by any multiple of 2. Once we make $\chi(\zeta) = 0$, it follows that $\partial X \cdot X = 0$. If $\ell \equiv 1(2)$, then $\chi(\zeta)$ is automatically 0 whatever the section we choose. By varying the section we can change $\delta X \cdot X$ by any multiple of k. We choose the section that makes $\delta X \cdot X = 0$. Now that both $\chi(\zeta)$ and $\delta X \cdot X$ are zero, we try to extend the immersion of δX over all of X. It is more convenient to view the problem as a relative problem for a manifold with boundary. Let $\bar{X}$ be X cut open along the k-sheets of δX. We have reduced $(\nu_{\bar{X}} - \psi^* \nu_L)$ to an ℓ-plane bundle over $\partial\bar{X}$. To complete the immersion over all of $\bar{X}$ it is necessary and sufficient to extend this bundle reduction over all of $\bar{X}$. The obstruction to doing this, if $\ell \equiv 0(2)$, is $\chi(\zeta)$. Thus such an extension exists in the case. If $\ell \equiv 1(2)$ then the obstruction is an element in $\mathbb{Z}/2$. It is calculated by taking an $(\ell+1)$ plane reduction of $(\nu_{\bar{X}} - \psi^* \nu_L)$, and taking the obstruction to extending the section given in the boundary over all of $\bar{X}$. This latter obstruction is an integer, and we are interested only in its residue class modulo 2. Such a bundle reduction corresponds to an immersion of X into $L \times I$ extending the given embedding of $\delta X \hookrightarrow L \times \{0\}$. The obstruction to extending the section is equal modulo 2 to $[X] \cdot [X]$. But all such homological intersections are 0 in $L \times I$. Thus the extended immersion of X into L exists. Since $\delta X \cdot X = 0$ we can deform X by regular homotopy until $\delta X \cap (X - \delta X) = \emptyset$. Then all self-intersections will be circles. If $2k + 1 \equiv 3(4)$, then all these circles must have preimages which are two circles. (The reason is that $\pi_1(L) = \{e\}$ and thus does not contain any elements, s, with $w_1(s) = (-1)$.) If $2k + 1 \equiv 1(4)$, then there can be such doubly covered circles. Let $S^{2r+1} \looparrowright \mathbb{R}^{4r+1}$ be any immersion whose normal bundle is the complement to a section in $\tau_{S^{2r+1}}$.

If we stabilize the immersion by adding one factor of $\mathbb{R}^1$ to the range, then the immersion, when shifted into general position, has normal bundle $\tau_{S^{2r+1}}$ and hence has an odd number of double points. By the argument in IV.3.2 we see that the original immersion $S^{2r+1} \looparrowright \mathbb{R}^{4r+1}$ must have had an odd number of circles of self-intersection whose preimage in S^{2r+1} was a single circle. By taking connected sum with such an immersion

we can change the number of such circles in $X^{k+1} \looparrowright L^{2k+1}$ ($k = 2r$) until it
is even. Once the number of these circles is even we can cancel them in
pairs as in IV.3.3.

<u>Corollary</u> IV.3.7: Let $T \hookrightarrow$ Tor $H_\ell(L^{2\ell+1})$ be a subkernel and $\{t_1, \ldots, t_r\}$ be
a minimal generating set for the two torsion subgroup of T. There are
immersed $\mathbb{Z}/n_i$-manifolds as in 3.6 $X_1^{\ell+1}, \ldots, X_r^{\ell+1} \looparrowright L^{2\ell+1}$ with δX_i representing t_i and with $\delta X_i \cdot X_j = \emptyset$.

<u>Proof</u>: By IV.3.6 we find the immersed $\{X_i^{\ell+1}\}$. Since the linking pairing
restricted to T is zero, we can deform the X_i until $\delta X_i \cap X_j = \emptyset, i \neq j$.

<u>Proposition</u> IV.3.8: There is a normal bordism $F: W \to \partial D \times L^{2\ell+1}$ from
$f \times 1_L: \partial U \times L \to \partial D \times L$ to a simple homotopy equivalence so that

   1) $K_*(W, \partial) = 0$ for $* \geq n + \ell + 1$,

   2) $K_*(W, \partial) \xrightarrow{\partial} K_{*-1}(\partial U \times L)$ is an isomorphism for $* \leq n + \ell - 1$,

   3) $K_{n+\ell}(W, \partial) \to K_{n+\ell-1}(\partial U \times L)$ is an isomorphism onto

       $K_\partial \otimes (H_\ell(L)/\text{Tor} \oplus T)$.

<u>Proof</u>: According to IV.3.4 we need only find $S \subset K_{n+\ell}(\partial U \times L; \mathbb{Q}/\mathbb{Z})$ so
that $\partial: S \xrightarrow{\cong} K_\partial \otimes T$ and $\mu|S = 0$. Since the image of μ is 2 torsion any
lifting of $K_\partial \otimes$ (odd torsion) will suffice.

     For the lifting on the 2-torsion we pick $\mathbb{Z}/n_i$-manifolds as in IV.3.7.
These crossed with a geometric basis for K_∂ for a generating set for an
appropriate S. The product of the $S^{n-1} \hookrightarrow \partial U$ with the immersion
$X_i^{\ell+1} \looparrowright L^{2\ell+1}$ is an immersion compatible with the bundle data and thus
appropriate for calculating the self-intersection function. The product
immersion is regularly homotopic to an embedding. This uses the fact
that $S^{n-1} \hookrightarrow \partial U$ has a normal field. (In fact its normal bundle is trivialized.) For a circle of self-intersections of $X_i^{\ell+1}$ in $L^{2\ell+1}$ shift the map
slightly in the direction of this normal field in the ∂U near one of its

preimage circles in X_i.

Now form $M_0 \times L \times I \cup W \times I \overset{G}{\to} N_0 \times L \times I$

Let g be the result of this surgery: $g: M_0 \times L \cup W \to N_0 \times L$. The exact
kernel sequence of the pair shows that

$$K_*(g) = \begin{cases} K_R \otimes H_{*-n}(L) & * \leq n + \ell - 2 \\[2ex] K_A \otimes H_{*-n+1}(L) & * \geq n + \ell + 1 \end{cases}$$

$$0 \longrightarrow K_A \otimes T^* \longrightarrow K_{n+\ell-1}(g) \longrightarrow K_R \otimes H_{\ell+1}(L) \longrightarrow 0$$

$$0 \longrightarrow K_A \otimes H_{\ell+1}(L) \longrightarrow K_{n+\ell}(g) \longrightarrow K_R \otimes (H_\ell(L)/\text{Tor} \oplus T) \longrightarrow 0.$$

The classes of the form $K_A \otimes H_*(L)$ are represented by product cycles in
$M_0 \times L$; the ones of the form $K_R \otimes H_*(L)$ are represented by relative product cycles in $M_0 \times L$ union with relative cycles in W. Both the last
two sequences are split. All this implies that the pairings remain the
obvious tensor product pairings, and thus the map is an s-nice normal
map. Clearly, $K_A \otimes H_{\ell+1}/\text{Tor} \hookrightarrow K_{n+\ell}(g)/\text{Tor}$ is a subspace on which the
intersection pairing vanishes. If the quadratic function vanishes also,
then it is a subkernel. To calculate the value of the quadratic function
on $(x \otimes y)$ we represent x by $S^{n-1} \hookrightarrow M_0$ and y (or some odd multiple)
by $Y^{\ell+1} \hookrightarrow L^{2\ell+1}$. Since the product of these two immersions is regularly
homotopic to an embedding, $q(x \otimes y) = 0$. Consequently, $K_A \otimes H_{\ell+1}/\text{Tor}$ is
a subkernel.

Before we can apply II.3.1 we must know that the quadratic function vanishes on the torsion submodule of $K_{n+\ell}(g)$. For $K_A \otimes \mathrm{Tor}\ H_{\ell+1}(L)$ this follows from the argument that showed $q|K_A \otimes H_{\ell+1}(L)/\mathrm{Tor}$ is zero. The submodule $K_R \otimes T$ is somewhat more delicate. Of course we need only consider 2 torsion classes. Let $t_1, \ldots, t_r \in T$ be the subset of the generators of T which have order a power of 2. Then in $L^{2\ell+1}$ we have immersed $\mathbb{Z}/n_i$-manifolds $X_1^{\ell+1}, \ldots, X_s^{\ell+1}$ whose Bocksteins represent the t_i. Let $\delta X_i'$ be a copy of δX_i pushed out along the normal field. Let $\{S_1^{n-1}, \ldots, S_{2r}^{n-1}\}$ be the geometric basis for $K_{n-1}(\partial U)$. Here the intersections of the S_i are all empty except for $S_{2i-1}^{n-1} \cap S_{2i}^{n-1}$, $i = 1, \ldots, r$, which is one point of transverse intersection. We have a family of disjointly embedded cycles $\{S_{2i-1}^{n-1} \times \delta X_j, \ S_{2i}^n \times \delta X_j'\}_{i=1, j=1}^{r \quad s}$. They represent the part of the natural generating set for the subkernel $K_{n-1}(\partial U) \otimes T \hookrightarrow \mathrm{Tor}\ K_{n+\ell-1}(f|\partial U \times 1_L)$ which is of order a power of 2. By IV.3.5 they bound disjointly embedded manifolds $Z_{i,j}^{n+\ell}$ in $W^{2n+2\ell}$. The normal fields over $S_{2i-1} \times \delta X_j$ and $S_{2i} \times \delta X_j'$ extend over the $Z_{i,j}$.

A representative for a class in $K_{n+\ell}(M_0 \times L \cup W)$ which projects to $\alpha \otimes t_j$ in $K_R \otimes \mathrm{Tor}\ H_\ell(L)$ is given by the following construction. Suppose $\partial \alpha = \Sigma \lambda_i [S_i]$. Pick an immersed manifold $V_\alpha^n \rightarrowtail M_0^{2n-1}$ whose boundary is geometrically $\Sigma \lambda_i S_i$. Then $V_\alpha \times \delta X_j \rightarrowtail M_0 \times L$ represents $\alpha \otimes t_i \in K_R \otimes T$. The self-intersections of this immersion are the self intersections of V_α crossed with δX_j. The self-intersections of V_α are one manifolds. Since $\delta X_j^\ell \hookrightarrow L^{2\ell+1}$ has two linearly independent normal fields we can remove all the self-intersections by deformation along these normal fields in $L^{2\ell+1}$. The boundary of the immersion after it is shifted to be an embedding is a linear combination

$$\sum \lambda_i S_i \times \delta X_j$$

where the various copies of δX_j have been shifted along a normal field for δX_j so that they are all disjoint. Each individual $S_i \times \delta X_j$ bounds $Z_{i,j}$ in W. If we have several parallel copies of $S_i \times \delta X_j$ in

$\partial(V_\alpha \times \partial X_j)$ then they will bound parallel copies of $Z_{i,j}$ (since the normal field $S_i \times \partial X_j$ extends over $Z_{i,j}$). The union

$$V_\alpha \times \partial X_j \cup \sum_i \lambda_i Z_{i,j}$$

thus is an immersed cycle in $M_0 \times L \cup W$ which represents an element in Tor $K_{n+\ell}(f \times 1_L)$ which projects to $\alpha \otimes t_j \in K_R \otimes T$. As we have noted this union is a correct immersion for calculating the value of the quadratic form. The above argument shows that it is regularly homotopic to an embedding. Thus the quadratic form vanishes on this element. Since these elements are in Tor $K_{n+\ell}(g)$, the intersection of any two of them is 0. Thus the quadratic form vanishes on a torsion module that classes of this type generate. Since we have already seen that the form vanishes on $K_A \otimes$ Tor $H_{\ell+1}(L^{2\ell+1})$ it follows that it vanishes in all of Tor $K_{n+\ell}(g)$.

Applying II.3.1 we see that we can perform surgery on

$$g: M_0 \times L \cup W \longrightarrow N_0 \times L$$

relative to its boundary to make it a simple homotopy equivalence. We are left to do surgery on the other side

$$W \cup U \times L \longrightarrow D \times L$$

relative to its boundary. The obstruction to doing this is an index (if $n + \ell \equiv 0(2)$) or a Kervaire obstruction (if $n + \ell \equiv 1(2)$). It agrees with the index or Kervarie obstruction of the original product $f \times 1_L: M^{2n-1} \times L^{2\ell+1} \to N^{2n-1} \times L^{2\ell+1}$. By the product formula for these simply connected obstructions, this obstruction is zero. Thus we can do surgery on $W \cup U \times L \to D \times L$ relative to its boundary to make the map a simple homotopy equivalence. Putting these two normal bordisms together gives one form $f \times 1_L: M \times L \to N \times L$ to a simple homotopy equivalence. Thus $\sigma(f \times 1_{L^{2\ell+1}}) = 0$ if $d(L^{2\ell+1}) = 0$. The general result, IV.1.1, then follows for any product by the same additivity argument given at the end

of section IV.1.

CHAPTER V: An Example

The missing information in the product formula in chapter 4 is a calculation of the maps $\varphi: L_n(\pi) \to L_{n+1}(\pi)$ and $\varphi^s: L_n^s(\pi) \to L_{n+1}^s(\pi)$. All we have shown about this map is that every element in its image is of order 1 or 2. In this section we will prove that φ is not always zero. In fact we will show that $\varphi: L_3(\mathbb{Z}, -) \to L_4(\mathbb{Z}, -)$ is an isomorphism $\varphi: \mathbb{Z}/2 \overset{\cong}{\to} \mathbb{Z}/2$. By crossing with S^1 one can produce, from this example, examples for every n where $\varphi: L_n(\pi) \to L_{n+1}(\pi)$ is non-zero, (or $\varphi^s: L_n^s(\pi) \to L_{n+1}^s(\pi)$ is non zero). However, all examples we know where φ is non-zero are derived from this one. For instance, we know no orientable example where $\varphi \neq 0$. The example we give here is just a reinterpretation of a result in [9] about $\mathbb{Z}/2$-manifolds in the language of non-simply connected surgery.

First we describe the groups $L_3(\mathbb{Z}, -)$ and $L_4(\mathbb{Z}, -)$, and show how to determine the surgery obstruction of a normal map between such manifolds. A $(\mathbb{Z}, -)$ manifold is a manifold M^N which admits a simply connected, closed submanifold $(\delta M)^{n-1}$ such that the normal bundle of δM^{n-1} in M^n is trivial and such that $M^n - (\delta M \times (-1,1))$ is a simply connected oriented manifold with oriented boundary $\delta M \cup \delta M$. It follows that $\pi_1(M) = \mathbb{Z}$ and that the generator reverses the orientation.

Let $f: M^{4n+3} \to N^{4n+3}$ be a degree one normal map with N a $(\mathbb{Z}, -)$ manifold. Put f transverse to $\delta N \hookrightarrow N$ and get the restricted map $f|: \delta M \to \delta N$. This normal map has a Kervaire invariant, [2], in $\mathbb{Z}/2$ which by a relative version of the above construction is seen to be an invariant of the normal bordism class of the original $(\mathbb{Z}, -)$ normal map. Suppose it vanishes. We can then assume $f|\delta M$ is a homotopy equivalence. Let $\bar{f}: \bar{M} \to \bar{N}$ be the normal map obtained by "opening up" f along δM and δN. It is a

normal map between simply connected manifolds which is a homotopy equivalence on the boundary.   Since the dimension of $\bar{N}$ is congruent to
3 modulo 4 we can perform surgery on f relative to $f|\delta\bar{M}$ to produce a
homotopy equivalence of pairs. This proves that if the Kervaire invariant
of f along δN is zero, then f is normal bordant to a homotopy equivalence. Conversely, if f: $M^{4n+3} \to N^{4n+3}$ is a homotopy equivalence between
$(\mathbb{Z}, -)$ manifolds, then one uses codimension 1 surgery techniques see [4]
and [18] to prove that it is deformable to a homotopy equivalence of
pairs f: $(M, \delta M) \to (N, \delta N)$. Thus if $\sigma(f) = 0$, then the Kervaire obstruction
of f along δN is zero.

This gives an injection $L_3(\mathbb{Z}, -) \to \mathbb{Z}/2$. To see that it is onto let
$K^{4k+2} \to S^{4k+2}$ be the basic normal map of Kervaire invariant 1, see [6].
K^{4k+2} is the plumbing of two tangent disk bundles of S^{2k+1} whose boundary
is coned off

$$\tau^{2k+1} \# \tau^{2k+1} \cup \text{cone (boundary)}$$

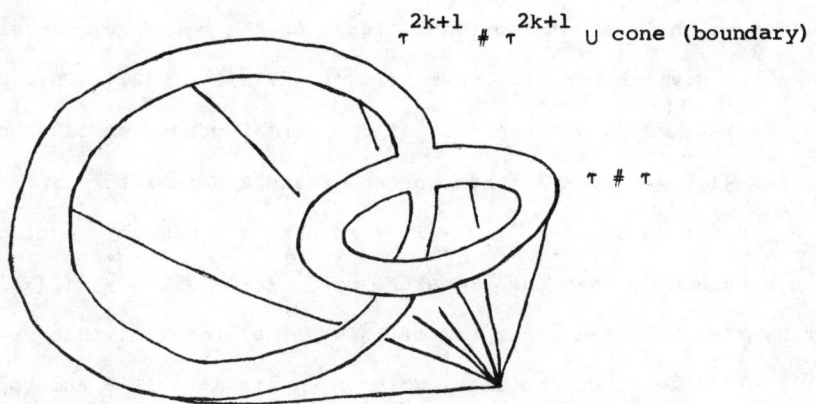

$\tau \# \tau$

The normal map ρ admits an orientation reversing homeomorphism h. The
map h switches the two copies of τ^{2k+1} and is extended by coning over
the cone on the boundary. On the sphere h is the suspension of the
homeomorphism induced by h on $\partial(\tau \# \tau)$ (which is an S^{4k+1}). We form

$$K \times I/h \xrightarrow{\ \rho \times I/h\ } S \times I/h.$$

This is a degree one normal map of $\mathbb{Z}/2$-manifolds whose $\mathbb{Z}/2$ obstruction is 1.

Let $g: X^{4k} \to Y^{4k}$ be a degree one normal map between $(\mathbb{Z}, -)$ manifolds. Put g transversal to $\delta Y \hookrightarrow Y$. Since δY^{4k-1} is simply connected we can do surgery on $g^{-1}(\delta Y) \xrightarrow{\ g|\ } \delta Y$ to produce a homotopy equivalence.

Let $\bar{g}: \bar{X} \to \bar{Y}$ be the "opened up" normal map after we have made g $g|g^{-1}(\delta Y) \to \delta Y$ a homotopy equivalence. Since $\bar{g}$ is a homotopy on the boundary, there is an integral obstruction $\frac{1}{8}[I(\bar{X}) - I(\bar{Y})]$, to finding a normal bordism relative to $\delta \bar{X}$ from $\bar{g}$ to a homotopy equivalence. Let $\sigma(g) \in \mathbb{Z}/2$ be $\frac{1}{8}[I(\bar{X}) - I(\bar{Y})]$ reduced modulo 2. Suppose $h: X' \to Y$ is another normal map with $h|h^{-1}(\delta Y) \to \delta Y$ a homotopy equivalence and $\bar{h}: \bar{X}' \to \bar{Y}$ the "opened up" map. If g and h are normally bordant by $H: W \to Y \times I$, then shift H relative to δW to be transverse to $\delta Y \times I$, and let Z^{4k} be $H^{-1}(\delta Y \times I)$. Clearly $\bar{W}$ is a bordism relative to the boundary from $\bar{X}$ to $Z \cup \bar{X}' \cup Z$. Since $Z^{4k} \to \delta Y \times I$ is a degree one normal map which is a homotopy equivalence on the boundary, the signature of Z is divisible by 8. Thus $\frac{1}{8}(I(\bar{X}') - I(\bar{X})) \equiv 0(2)$. This proves that $\sigma(g) \in \mathbb{Z}/2$ is an invariant of the normal cobordism class of g.

If $\sigma(g) = 0$ in $\mathbb{Z}/2$, then we can create a normal bordism from g $H: W \to Y \times I$ so that $H^{-1}(\delta Y \times I) \to \delta Y \times I$ is a homotopy equivalence on both ends and so that the signature of $H^{-1}(\delta Y \times I)$ is $-\frac{1}{2}(I(\bar{X}) - I(\bar{Y}))$. Let $h: X' \to Y$ be the other end of this normal bordism. Since $I(\bar{X}') - I(\bar{Y}) = 0$, we can do surgery on h to make it a homotopy equivalence.

Conversely, if g is normally bordant to a homotopy equivalence, then by codimension 1 techniques, [18], we can make g a homotopy equivalence of pairs, and thus this $\mathbb{Z}/2$-invariant $\sigma(g)$ is 0.

The main theorem that we need from [9] is the following.

__Lemma__ ([9] - theorem 6.1 Case 1): If $f: M^{4k+2} \to N^{4k+2}$ is a normal map between closed simply connected manifolds and $L^{4\ell+1}$ is a closed manifold, then the index of any normal bordism of $f \times 1_L : M \times L \to N \times L$ to a homotopy equivalence is $4 \cdot \sigma(f) \cdot d(L)$ in $\mathbb{Z}/8\mathbb{Z}$. Here, $\sigma(f)$ is the Kervaire obstruction of f in $\mathbb{Z}/2$, and $d(L)$ is the de Rham invariant of L in $\mathbb{Z}/2$.

__Theorem__ V.1: $\varphi^s : L_3^s(\mathbb{Z}, -) \to L_4^s(\mathbb{Z}, -)$ is an isomorphism.

__Proof__: Take $f : M^{4k+3} \to N^{4k+3}$ to be any normal map representing the non-zero element in $L_3^s(\mathbb{Z}, -)$. To calculate φ^s, we must cross with any closed, simply connected manifold L^5 with $d(L^5) = 1$. First put f transverse to δN, and call δM the preimage. Opening up along δN and δM gives $\bar{f} : \bar{M} \to \bar{N}$. Crossing with L^5 gives $\bar{f} \times 1_L : \bar{M} \times L \to \bar{N} \times L$. This map is not a homotopy equivalence on the boundary. If W is any normal bordism from $\delta M \times L \to \delta M \times L$ to a homotopy equivalence, then $I(W) \equiv 4(8)$. According to our description of $L_4^s(\mathbb{Z}, -)$,
$\sigma(f \times 1_L) = \{\frac{1}{8}[I(W \cup \bar{M} \times L \cup W) - I(\bar{N} \times L)]\} \bmod 2$. By the Novikov additivity formula for the index and the fact that $I(\bar{M} \times L) = I(\bar{N} \times L)$ $= 0$, we have $\sigma(f \times 1_L) = \{\frac{1}{8}[2 \, I(\bar{W})]\} \bmod 2 = \{\frac{1}{8}[2.4]\} \bmod 2 = 1 \bmod 2$.

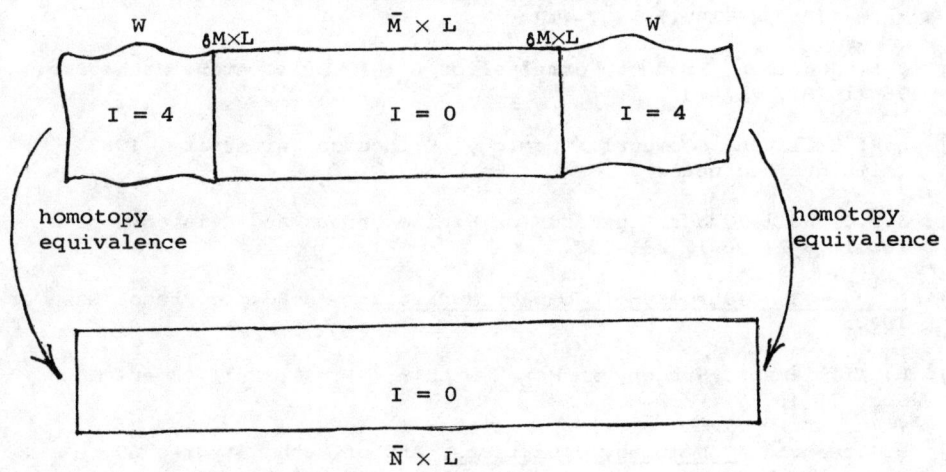

This proves φ^s is an isomorphism.

REFERENCES

[1]   M.F. Atiyah, Thom Complexes, Proc. London Math. Soc. (3) 11 (1961), 291-310.

[2]   W. Browder, Surgery on Simply-Connected Manifolds, Springer-Verlag, New York, 1972.

[3]   P.E. Conner and E.E. Floyd, Differentiable Periodic Maps, Academic Press, New York, 1964.

[4]   F.T. Farrell, The Obstruction to Fibering a Manifold Over a Manifold Over a Circle, Bull. Amer. Math. Soc. 73 (1967), 737-740.

[5]   M.W. Hirsch, Immersions of Manifolds, Trans. Amer. Math. Soc. 93 (1959), 242-276.

[6]   M.A. Kervaire and J.W. Milnor, Groups of Homotopy Spheres: I, Ann. of Math. 77 (1963), 504-537.

[7]   J.W. Milnor, On Simply Connected 4-manifolds, Internat. Sympos. Algebraic Topology, Univ. Nacunal Autonoma de Mexico, 1956, 122-128.

[8]   R.J. Milgram, Surgery with Coefficients, Ann. of Math. 100 (1974), 194-248.

[9]   J.W. Morgan and D.P. Sullivan, The Transversality Characteristic Class and Linking Cycles in Surgery Theory, Ann. of Math. 99 (1974), 463-544.

[10] S.P. Novikov, Homotopically Equivalent Smooth Manifolds I, Translations Amer. Math. Soc. 48, 271-396.

[11] C.P. Rourke and D.P. Sullivan, On the Kervaire Obstruction, Ann. of Math. (2) 94 (1971), 397-413.

[12] J.L. Shaneson, Product Formulas for $L_n(\pi)$, Bull. Amer. Math. Soc. 76 (1970), 787-791.

[13] D.P. Sullivan, Geometric Topology, Princeton University, 1967 (mimeographed notes).

[14] C.T.C. Wall, Quadratic Forms on Finite Groups and Related Topics, Topology 2 (1963), 281-298.

[15] _____, Surgery of Compact Manifolds, Academic Press, New York, 1970.

[16] R. Williamson, Surgery in M × N with $\pi_1(M) \neq 1$, Bull. Amer. Math. Soc. 75 (1969), 582-585.

[17] N. Steenrod, Cohomology Operations, Ann of Math. Studies 50, Princeton University Press, Princeton, N.J., 1962.

[18] J. Shaneson, Wall's Surgery obstruction groups for $\mathbb{Z} \times G$, Ann. of Math. 90 (1969), 296-334.

Department of Mathematics, Columbia University, New York, N. Y. 10027